T0122886

Introduction to Differential Equations Using Sage

Introduction to Differential Equations Using Sage

David Joyner and Marshall Hampton

The Johns Hopkins University Press
Baltimore

Printed in the United States of America on acid-free paper
9 8 7 6 5 4 3 2 1

The Johns Hopkins University Press
2715 North Charles Street
Baltimore, Maryland 21218-4363
www.press.jhu.edu

ISBN 13: 978-1-4214-0637-4 (hc)
ISBN 10: 1-4214-0637-3 (hc)
ISBN 13: 978-1-4214-0724-1 (ebook)
ISBN 10: 1-4214-0724-8 (ebook)

Library of Congress Library Control Number: 2012934439

A catalog record for this book is available from the British Library.

The Johns Hopkins University Press uses environmentally friendly book materials, including recycled text paper that is composed of at least 30 percent post-consumer waste, whenever possible.

There are some things which cannot
be learned quickly, and time, which is all we have,
must be paid heavily for their acquiring.
They are the very simplest things,
and because it takes a man's life to know them
the little new that each man gets from life
is very costly and the only heritage he has to leave.

—Ernest Hemingway, *from A. E. Hotchner*, Papa Hemingway

Contents

Preface

The majority of this book came from lecture notes David Joyner typed up over the years for a course on differential equations with boundary value problems at the U. S. Naval Academy (USNA). After Joyner finished a draft of this book, he invited the second author, Marshall Hampton, to revise and extend it. At the University of Minnesota Duluth, Hampton teaches a course on differential equations and linear algebra. The differential equations course at the USNA has used various editions of the following three books (in order of most common use to least common use) at various times:

- Dennis G. Zill and Michael R. Cullen, *Differential Equations with Boundary Value Problems*, 6th ed. (Belmont, CA: Brooks/Cole, 2005).
- R. Nagle, E. Saff, and A. Snider, *Fundamentals of Differential Equations and Boundary Value Problems*, 4th ed. (Reading, MA: Addison-Wesley, 2003).
- W. Boyce and R. DiPrima, *Elementary Differential Equations and Boundary Value Problems*, 8th ed. (New York: John Wiley and Sons, 2005).

You may see some similarities but, for the most part, we taught things a bit differently and tried to impart a more computational perspective in these example-based course notes.

A new feature to this book is the fact that every section has at least one `Sage` exercise. `Sage` is FOSS (free and open source software), available on the most common computer platforms. Most examples are also illustrated using `Sage` commands. One section also uses `SymPy`, another FOSS computer algebra system. Some exercises in the text are considerably more challenging than others. We have indicated these with an asterisk next to the exercise number.

The book starts off with first-order differential equations, discussing methods of solution, the existence and uniqueness of solutions, and common applications to physics and electrical engineering. We also show how to plot approximations to solutions using numerical methods (such as Euler's method) and the direction field of a first-order ordinary differential equation.

Higher-order linear differential equations are discussed next. We solve nonhomogeneous linear differential equations by various methods, such as undetermined coefficients, annihilators, power series, and Laplace transforms. Most of the emphasis is on constant-coefficient linear differential equations. We study applications to vibrating spring problems and electrical circuit problems in detail. In contrast to the texts mentioned above, we present details on the impulse-response formula for the solution to a constant-coefficient linear differential equation in terms of the weight function convolved with the inverse Laplace transform of the forcing function.

The third chapter discusses systems of differential equations and their background in matrix theory. We show how to use the row-reduced echelon form to solve linear systems and to compute the inverse of the matrix, and how to compute the eigenvalues and eigenvectors of a matrix. These techniques are used to solve systems of differential equations by matrix (i.e., eigenvalue) methods and by Laplace transforms. We explain how to rewrite a system of higher-order equations as a system of first-order equations and how to solve it numerically (using, for example, Euler's method). Applications to Lanchester's equations and electric networks, among others, are presented. In particular, several electric network example problems are solved in detail using Laplace transform methods. This text presents in detail Lanchester's system of differential equations, which models a simple battle between two forces. The Battle of Trafalgar is illustrated using this model. We also show how Sage can be used to numerically solve a nonlinear system of differential equations which models a zombie attack.

The last chapter discusses partial differential equations (PDEs), with emphasis on specific PDEs having applications to physics. We show how to solve simple PDEs using the method of separation of variables. We start with the advection equation, which is a first-order linear PDE used to model traffic flow or the flow of silt in a river. In preparation for Fourier's method to solve the heat equation, we present background on the Fourier series expansion of a given periodic function on the real line. This background is applied to solve the heat equation for a wire, as well as the wave equation for a string, using Fourier's method. The chapter (and the book) ends with a brief discussion of the one-dimensional Schrödinger equation for a free particle in a box.

The authors donate the royalties of this book to the (501(c) nonprofit) Sage Foundation to further promote free mathematical software for everyone.

Acknowledgments

In a few cases we have made use of the *excellent* (public domain!) lecture notes by Sean Mauch, "Introduction to methods of Applied Mathematics," available online at `http://www.its.caltech.edu/~sean/book/unabridged.html` (as of early 2012).

In some cases, we have made use of the material on Wikipedia—this includes both discussion and in a few cases, diagrams or graphics. This material is licensed under the GFDL or the Attribution-ShareAlike Creative Commons license. In any case, the amount used here probably falls under the "fair use" clause.

Software used: Most of the graphics in this text were created using `Sage` (`http://www.sagemath.org/`) and `GIMP` (`http://www.gimp.org/`) by the authors. The most important components of `Sage` for our purposes are `Maxima`, `SymPy` and `Matplotlib`. The circuit diagrams were created using `Dia` (`http://www.gnome.org/projects/dia/`) and `GIMP` by the authors. A few diagrams were "taken" from Wikipedia (`http://www.wikipedia.org/`), and acknowledged in the appropriate place in the text. Of course, LaTeX was used for the typesetting, and `emacs` (`http://www.gnu.org/software/emacs/`) for editing. Many thanks to the developers of these programs, and especially the `Sage` developers, for these great free software tools.

Introduction to Differential Equations Using Sage

Chapter 1

First-order differential equations

> But there is another reason for the high repute of mathematics: it is mathematics that offers the exact natural sciences a certain measure of security which, without mathematics, they could not attain.
>
> —*Albert Einstein*

1.1 Introduction to DEs

Roughly speaking, a differential equation (DE) is an equation involving the derivatives of one or more unknown functions. Implicit in this vague definition is the assumption that the equation imposes a *constraint* on the unknown function (or functions). For example, we would not call the well-known product rule identity of differential calculus a differential equation.

Example 1.1.1. Here is a simple example of a differential equation: $\frac{dx}{dt} = x$. A solution would be a function $x = x(t)$ that is differentiable on some interval and whose derivative is equal to itself. Can you guess any solutions?

As far as notation goes, in this text we often drop the dependence on the independent variable when referring to a function. For example, $x(t)$ is sometimes simply denoted by x.

In calculus (differential, integral, and vector), you've studied ways of analyzing functions. You might even have been convinced that functions you meet in applications arise naturally from physical principles. As we shall see, differential equations arise *naturally* from general physical principles. In many cases, the functions you met in calculus in applications to physics were actually solutions to a naturally arising differential equation.

Example 1.1.2. Consider a falling body of mass m on which exactly three forces act:

- gravitation, F_{grav};
- air resistance, F_{res};
- an external force, $F_{ext} = f(t)$, where $f(t)$ is some given function.

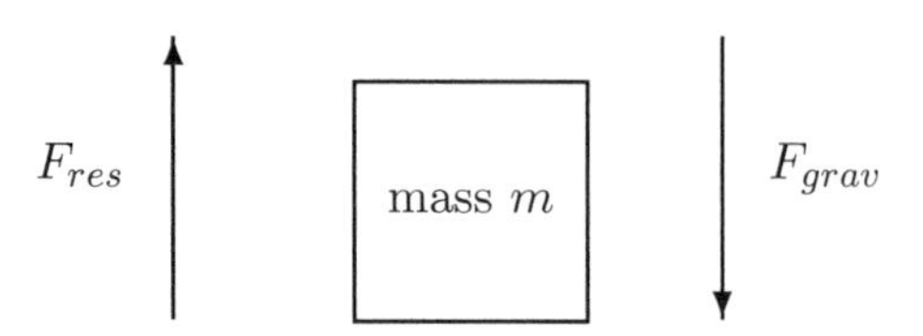

Let $x(t)$ denote the distance fallen from some fixed initial position. (In other words, we measure $x(t)$ so that *down is positive.*) The velocity is denoted by $v = x'$ and the acceleration by $a = x''$. In this case, gravitational force is governed by $F_{grav} = mg$, where $g > 0$ is the gravitational constant. We assume that air resistance F_{res} is proportional to velocity (a common assumption in physics) and write $F_{res} = -kv = -kx'$, where $k > 0$ is a "friction constant." The total force F_{total} is by hypothesis

$$F_{total} = F_{grav} + F_{res} + F_{ext},$$

and, by Newton's second law for a constant mass,

$$F_{total} = (mx')' = mx'' = ma.$$

Putting these together, we have

$$mx'' = ma = mg - kx' + f(t),$$

or

$$mx'' + kx' = f(t) + mg.$$

This is a differential equation in $x = x(t)$. It may also be rewritten as a differential equation in $v = v(t) = x'(t)$ as

$$mv' + kv = f(t) + mg.$$

This is an example of a "first-order differential equation in v," which means that at most first-order derivatives of the unknown function $v = v(t)$ occur.

In fact, you have probably seen solutions to this in your calculus classes, at least when $f(t) = 0$ and $k = 0$. In that case, $v'(t) = g$ and so $v(t) = \int g\, dt = gt + C$. Here the constant of integration C represents the initial velocity.

Example 1.1.3. Assume that if two armies fight, with $x(t)$ troops on one side and $y(t)$ on the other, then the rate at which soldiers in one army are put out of action is proportional to the troop strength of their enemy. This gives rise to the system of differential equations

$$\begin{cases} x'(t) = -Ay(t), & x(0) = x_0, \\ y'(t) = -Bx(t), & y(0) = y_0, \end{cases}$$

where $A > 0$ and $B > 0$ are constants and x_0 and y_0 are the initial troop strengths. This will be discussed in more detail in §3.3.1.

Differential equations occur in other areas as well: weather prediction (more generally, fluid-flow dynamics), electrical circuits, the temperature of a heated homogeneous wire, and many others (see the Table 1.1 below). They even arise in problems on Wall Street: the Black-Scholes equation is a partial differential equation which models the pricing of derivatives [BS-intro]. Learning to solve differential equations helps you understand the behaviour of phenomenon present in these problems.

Phenomenon	Description of DE
weather	Navier-Stokes equation [NS-intro] a nonlinear vector-valued higher-order PDE
falling body	first-order linear ODE
motion of a mass attached to a spring	Hooke's spring equation second-order linear ODE [H-intro]
motion of a plucked guitar string	Wave equation second-order linear PDE [W-intro]
Battle of Trafalgar	Lanchester's equations system of two first-order DEs [L-intro], [M-intro], [N-intro]
cooling cup of coffee in a room	Newton's Law of Cooling first-order linear ODE
population growth	logistic equation nonlinear, separable, first-order ODE

Table 1.1: Some applications

Undefined terms and notation will be defined below, except for the equations themselves. For those, see the references or wait until later sections when they will be introduced.[1]

Basic Concepts

[1]For example, the important Navier-Stokes equation is too complicated to state here and its explanation would take us too far afield. For details, see, for example, `http://en.wikipedia.org/wiki/Navier-stokes` or [A-ns].

Here are some of the concepts (and abbreviations) to be introduced:

- dependent variable(s),
- independent variable(s),
- ordinary differential equations (ODEs),
- partial differential equations (PDEs),
- order,
- linearity, and
- solution.

It is really best to learn these concepts using examples. However, here are the general definitions anyway, with examples to follow.

Notation and definitions

- The term "differential equation" is sometimes abbreviated DE.
- **Dependent and independent variables**. Put simply, a differential equation is an equation involving derivatives of one or more unknown functions. The variables you are differentiating with respect to are the *independent variables* of the DE. The variables (the "unknown functions") you are differentiating are the *dependent variables* of the DE. Other variables that might occur in the DE are sometimes called "parameters."
- **ODE and PDE**. If none of the derivatives that occur in the differential equation are partial derivatives (for example, if the dependent variable or unknown function is a function of a single variable), then the differential equation is called an *ordinary differential equation* or *ODE*. If some of the derivatives that occur in the differential equation are partial derivatives, then the differential equation is a *partial differential equation* or *PDE*.
- **Order**. The highest total number of derivatives you have to take in the differential equation is its *order*.
- **Linearity**. This can be described in a few different ways. First of all, a differential equation is *linear* if the only operations you perform on its terms are combinations of the following:

  — differentiation with respect to independent variable(s);

  — multiplication by a function of the independent variable(s).

Another way to define linearity is as follows. A *linear ODE* having independent variable t and dependent variable y is an ODE of the form

$$a_0(t)y^{(n)} + \cdots + a_{n-1}(t)y' + a_n(t)y = f(t),$$

for some given functions $a_0(t)$, ..., $a_n(t)$, and $f(t)$. Here

$$y^{(n)} = y^{(n)}(t) = \frac{d^n y(t)}{dt^n}$$

denotes the nth derivative of $y = y(t)$ with respect to t. The terms $a_0(t)$, ..., $a_n(t)$ are called the *coefficients* of the differential equation and we will call the term $f(t)$ the *nonhomogeneous term* or the *forcing function*. (In physical applications, this term usually represents an external force acting on the system. For instance, in the example above it represents the gravitational force mg.)

A *linear PDE* is a differential equation which is a linear ODE in each of its independent variables.

- **Solution**. An explicit *solution* to an ODE having independent variable t and dependent variable x is simply a function $x(t)$ for which the differential equation is true for all values of t in an open interval. An *implicit solution* to an ODE is an implicitly defined (possibly multiple-valued) function which satisfies the ODE.

- **IVP**. A first-order *initial value problem* (abbreviated *IVP*) is a problem of the form

$$x' = f(t, x), \quad x(a) = c, \tag{1.1}$$

where $f(t, x)$ is a given function of two variables, and a, c are given constants. The equation $x(a) = c$ is the *initial condition* (IC).

Here are some examples.

Example 1.1.4. Table 1.2 lists some examples. As an exercise, determine which of the following are ODEs and which are PDEs.

Remark 1.1.1. Note that in many of these examples, the symbol used for the independent variable is not made explicit. For example, we are writing x' when we really mean $x'(t) = \frac{dx(t)}{dt}$. This is very common shorthand notation and, in this situation, we shall use t as the independent variable whenever possible.

Example 1.1.5. Recall that a linear ODE having independent variable t and dependent variable y is an ODE of the form

$$a_0(t)y^{(n)} + \cdots + a_{n-1}(t)y' + a_n(t)y = f(t),$$

DE	Independent Variable(s)	Dependent Variable(s)	Order	Linear or Nonlinear?
$mx'' + kx' = mg$ falling body	t	x	2	yes
$mv' + kv = mg$ falling body	t	v	1	yes
$k\frac{\partial^2 u}{\partial x^2} = \frac{\partial u}{\partial t}$ heat equation	t, x	u	2	yes
$mx'' + bx' + kx = f(t)$ spring equation	t	x	2	yes
$P' = k(1 - \frac{P}{K})P$ logistic population equation	t	P	1	no
$k\frac{\partial^2 u}{\partial x^2} = \frac{\partial^2 u}{\partial^2 t}$ wave equation	t, x	u	2	yes
$T' = k(T - T_{room})$ Newton's Law of Cooling	t	T	1	yes
$x' = -Ay$, $y' = -Bx$, Lanchester's equations	t	x, y	1	yes

Table 1.2: Some examples

for some given functions $a_0(t)$, ..., $a_n(t)$, and $f(t)$. The order of this differential equation is n.

In particular, a linear first-order ODE having independent variable t and dependent variable y is an ODE of the form

$$a_0(t)y' + a_1(t)y = f(t),$$

for some $a_0(t)$, $a_1(t)$, and $f(t)$. We can divide both sides of this equation by the leading coefficient $a_0(t)$ without changing the solution y to this differential equation. Let's do that and rename the terms:

$$y' + p(t)y = q(t),$$

where $p(t) = a_1(t)/a_0(t)$ and $q(t) = f(t)/a_0(t)$. Every linear first-order ODE can be put into this form, for some p and q. For example, the falling body equation $mv' + kv = f(t) + mg$ has this form after dividing by m and renaming v as y.

What does a differential equation like $mx'' + kx' = mg$ or $P' = k(1 - \frac{P}{K})P$ or $k\frac{\partial^2 u}{\partial x^2} = \frac{\partial^2 u}{\partial^2 t}$ really mean? In $mx'' + kx' = mg$, m and k and g are given constants. The only things that can vary are t and the unknown function $x = x(t)$. Let's consider and even more concrete example.

Example 1.1.6. To be specific, let's consider $x' + x = 1$. This means *for all* t, $x'(t) + x(t) = 1$. In other words, a solution $x(t)$ is a function which, when added to its derivative *always* gives the constant 1. How many functions are there with that property? Try guessing a few "random" functions:

- Guess $x(t) = \sin(t)$. Compute $(\sin(t))' + \sin(t) = \cos(t) + \sin(t) = \sqrt{2}\sin(t + \frac{\pi}{4})$. $x'(t) + x(t) = 1$ is *false*.
- Guess $x(t) = \exp(t) = e^t$. Compute $(e^t)' + e^t = 2e^t$. $x'(t) + x(t) = 1$ is *false*.
- Guess $x(t) = t^2$. Compute $(t^2)' + t^2 = 2t + t^2$. $x'(t) + x(t) = 1$ is *false*.
- Guess $x(t) = \exp(-t) = e^{-t}$. Compute $(e^{-t})' + e^{-t} = 0$. $x'(t) + x(t) = 1$ is *false*.
- Guess $x(t) = 1$. Compute $(1)' + 1 = 0 + 1 = 1$. $x'(t) + x(t) = 1$ is *true*.

We finally found a solution by considering the constant function $x(t) = 1$. Here a way of doing this kind of computation with the aid of the computer algebra system Sage:

Sage

```
sage: t = var('t')
sage: de = lambda x: diff(x,t) + x - 1
sage: de(sin(t))
sin(t) + cos(t) - 1
sage: de(exp(t))
2*e^t - 1
```

```
sage: de(t^2)
t^2 + 2*t - 1
sage: de(exp(-t))
-1
sage: de(1)
0
```

In the Sage code above, we have rewritten $x' + x = 1$ as $x' + x - 1 = 0$ and then plugged in various functions ($\sin(t)$, $\exp(t) = e^t$, and so on) for x to see if we get 0 or not.

Obviously, we want a more systematic method for solving such equations than guessing all the types of functions we know one by one. We will get to those methods in time. First, we need some more terminology.

Under mild conditions on f, an IVP of the form (1.1) has a solution $x = x(t)$ which is unique. This means that if f and a are fixed but c is a parameter then the solution $x = x(t)$ will depend on c. This is stated more precisely in the following result.

Theorem 1.1.1. *(Existence and uniqueness)* Fix a point (t_0, x_0) in the plane. Let $f(t,x)$ be a function of t and x for which both $f(t,x)$ and $f_x(t,x) = \frac{\partial f(t,x)}{\partial x}$ are continuous on some rectangle

$$a < t < b, \quad c < x < d$$

in the plane. Here a, b, c, d are any numbers for which $a < t_0 < b$ and $c < x_0 < d$. Then there is an $h > 0$ and a unique solution $x = x(t)$ for which

$$x' = f(t,x) \quad \text{for all } t \in (t_0 - h, t_0 + h),$$

and $x(t_0) = x_0$.

We will return to it in more detail in §1.3. (See §2.8 of Boyce and DiPrima [BD-intro] for more details on a proof.) In most cases we shall run across, it is easier to construct the solution than to prove this general theorem!

Example 1.1.7. Let us try to solve

$$x' + x = 1, \quad x(0) = 1.$$

The solution to the differential equation $x' + x = 1$ which we "guessed at" in the previous example, $x(t) = 1$, satisfies this IVP.

Here is a way of finding this solution with the aid of the computer algebra system Sage:

Sage

```
sage: t = var('t')
sage: x = function('x', t)
sage: de = lambda y: diff(y,t) + y - 1
```

```
sage: desolve(de(x),[x,t],[0,1])
1
```

(The command **desolve** is a differential equation solver in Sage.) Just as an illustration, let's try another example. Let us try to solve

$$x' + x = 1, \quad x(0) = 2.$$

The Sage commands are similar:

Sage

```
sage: t = var('t')
sage: x = function('x', t)
sage: de = lambda y: diff(y,t) + y - 1
sage: x0 = 2 # this is for the IC x(0) = 2
sage: soln = desolve(de(x),[x,t],[0,x0])
sage: P = plot(soln,0,5)
sage: soln; show(P, ymin = 0)
(e^t + 1)*e^(-t)
```

This gives the solution $x(t) = (e^t + 1)e^{-t} = 1 + e^{-t}$ and the plot given in Figure 1.1.

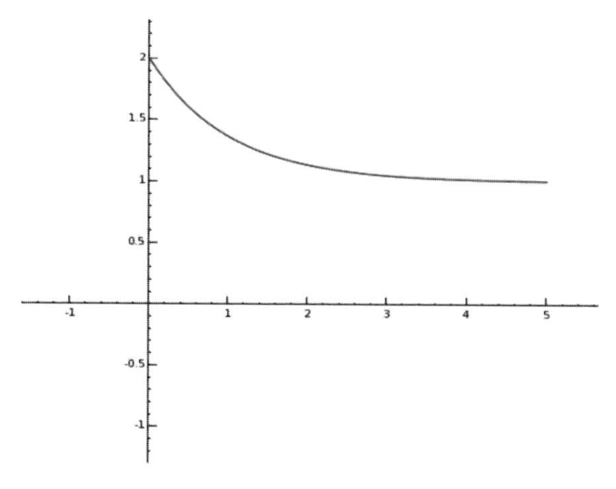

Figure 1.1: Solution to IVP $x' + x = 1$, $x(0) = 2$.

Example 1.1.8. Now let us consider an example that does not satisfy all of the assumptions of the existence and uniqueness theorem: $x' = x^{2/3}$. The function $x = (t/3+C)^3$ is a solution to this differential equation for any choice of the constant C. If we consider only solutions with the initial value $x(0) = 0$, we see that we can choose $C = 0$—i.e., $x = t^3/27$ satisfies the differential equation and has $x(0) = 0$. But there is another solution with that initial condition—the solution $x = 0$. So solutions exist, but they are not necessarily unique. This does not violate the existence and uniqueness theorem because in this case $f(x,t) = x^{2/3}$ is continuous near the initial condition $(x_0, t_0) = (0, 0)$ but its x-derivative $\frac{\partial f}{\partial x} = \frac{2}{3}x^{-1/3}$ is not continuous—indeed it is not even defined for $x = 0$.

Some conditions on the ODE which guarantee a unique solution will be presented in §1.3.

Exercises

1. Verify that $y = 2e^{-3x}$ is a solution to the differential equation $y' = -3y$.

2. Verify that $x = t^3 + 5$ is a solution to the differential equation $x' = 3t^2$.

3. Substitute $x = e^{rt}$ into the differential equation $x'' + x' - 6x = 0$ and determine all values of the parameter r that give a solution.

4. (a) Verify that, for any constant c, the function $x(t) = 1 + ce^{-t}$ solves $x' + x = 1$. Find the c for which this function solves the IVP $x' + x = 1$, $x(0) = 3$.

   (b) Solve

$$x' + x = 1, \quad x(0) = 3$$

   using Sage.

5. Write a differential equation for a population P that is changing in time (t) such that the rate of change is proportional to the square root of P.

1.2 Initial value problems

Recall that a first-order initial value problem, or IVP, is simply a first-order ODE and an initial condition. For example,

$$x'(t) + p(t)x(t) = q(t), \quad x(0) = x_0,$$

where $p(t)$, $q(t)$ and x_0 are given. The analog of this for second-order linear differential equations is this:

$$a(t)x''(t) + b(t)x'(t) + c(t)x(t) = f(t), \quad x(0) = x_0, \ x'(0) = v_0,$$

where $a(t)$, $b(t)$, $c(t)$, x_0, and v_0 are given. This second-order linear differential equation and initial conditions is an example of a second-order IVP. In general, in an IVP, the number of initial conditions must match the order of the differential equation.

Example 1.2.1. Consider the second-order differential equation

$$x'' + x = 0.$$

(We shall run across this differential equation many times later. As we will see, it represents the displacement of an undamped spring with a mass attached.) Suppose we know that the general solution to this differential equation is

$$x(t) = c_1 \cos(t) + c_2 \sin(t)$$

for any constants c_1, c_2. This means every solution to the differential equation must be of this form. (If you don't believe this, you can at least check if it is a solution by computing $x''(t)+x(t)$ and verifying that the terms cancel, as in the following Sage example. Later, we see how to derive this solution.) Note that there are two degrees of freedom (the constants c_1 and c_2), matching the order of the differential equation.

Sage

```
sage: t = var('t')
sage: c1 = var('c1')
sage: c2 = var('c2')
sage: de = lambda x: diff(x,t,t) + x
sage: de(c1*cos(t) + c2*sin(t))
0
sage: x = function('x', t)
sage: soln = desolve(de(x),[x,t]); soln
k1*sin(t) + k2*cos(t)
sage: solnx = lambda s: RR(soln.subs(k1=1, k2=0, t=s))
sage: P = plot(solnx,0,2*pi)
sage: show(P)
```

This solution is displayed in Figure 1.2.

Now, to solve the IVP

$$x'' + x = 0, \quad x(0) = 0, \ x'(0) = 1,$$

the problem is to solve for c_1 and c_2 for which the $x(t)$ satisfies the initial conditions. There are two degrees of freedom in the general solution, matching the number of initial conditions in the IVP. Plugging $t = 0$ into $x(t)$ and $x'(t)$, we obtain

$$0 = x(0) = c_1 \cos(0) + c_2 \sin(0) = c_1, \quad 1 = x'(0) = -c_1 \sin(0) + c_2 \cos(0) = c_2.$$

Therefore, $c_1 = 0$, $c_2 = 1$, and $x(t) = \sin(t)$ is the unique solution to the IVP.

In Figure 1.2, you see that the solution oscillates, as t increases.

Here is another example.

Example 1.2.2. Consider the second-order differential equation

$$x'' + 4x' + 4x = 0.$$

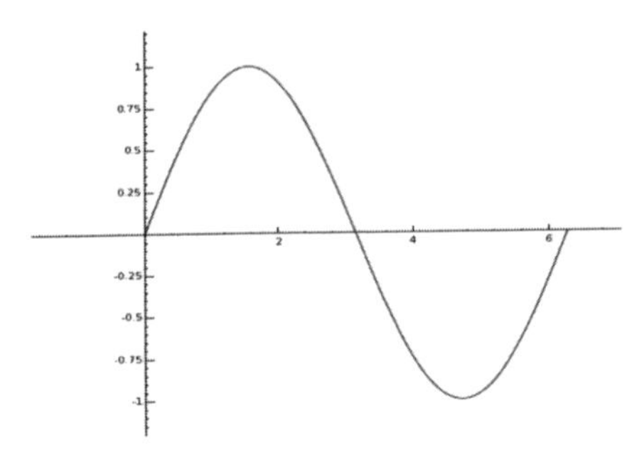

Figure 1.2: Solution to IVP $x'' + x = 0$, $x(0) = 0$, $x'(0) = 1$.

(We shall run across this differential equation many times later as well. As we will see, it represents the displacement of a critially damped spring with a unit mass attached.) Suppose we know that the general solution to this differential equation is

$$x(t) = c_1 \exp(-2t) + c_2 t \exp(-2t) = c_1 e^{-2t} + c_2 t e^{-2t}$$

for any constants c_1, c_2. This means that every solution to the differential equation must be of this form. (Again, you can at least check if it is a solution by computing $x''(t)$, $4x'(t)$, and $4x(t)$, adding them up and verifying that the terms cancel, as in the following Sage example.)

Sage

```
sage: t = var('t')
sage: c1 = var('c1')
sage: c2 = var('c2')
sage: de = lambda x: diff(x,t,t) + 4*diff(x,t) + 4*x
sage: de(c1*exp(-2*t) + c2*t*exp(-2*t))
0
sage: desolve(de(x),[x,t])
(k2*t + k1)*e^(-2*t)
sage: P = plot(t*exp(-2*t),0,pi)
sage: show(P)
```

The plot is displayed in Figure 1.3.

Now, to solve the IVP

$$x'' + 4x' + 4x = 0, \quad x(0) = 0,\ x'(0) = 1,$$

we solve for c_1 and c_2 using the initial conditions. Plugging $t = 0$ into $x(t)$ and $x'(t)$, we obtain

$$0 = x(0) = c_1 \exp(0) + c_2 \cdot 0 \cdot \exp(0) = c_1,$$

$$1 = x'(0) = c_1 \exp(0) + c_2 \exp(0) - 2c_2 \cdot 0 \cdot \exp(0) = c_1 + c_2.$$

Therefore, $c_1 = 0$, $c_1 + c_2 = 1$, and so $x(t) = t\exp(-2t)$ is the unique solution to the IVP. In Figure 1.3, you see that the solution tends to 0 as t increases.

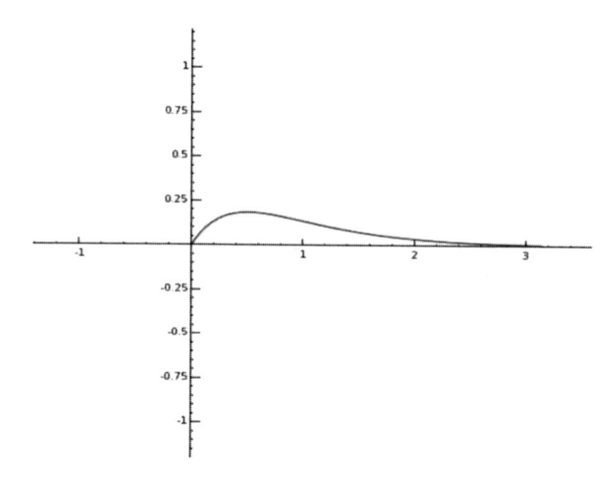

Figure 1.3: Solution to IVP $x'' + 4x' + 4x = 0$, $x(0) = 0$, $x'(0) = 1$.

Suppose, for the moment, that for some reason you mistakenly thought that the general solution to this differential equation was

$$x(t) = c_1 \exp(-2t) + c_2 \exp(-2t) = e^{-2t}(c_1 + c_2)$$

for arbitrary constants c_1, c_2. (Note: the "extra t-factor" on the second term on the right is missing from this expression.) *Now*, if you try to solve for the constants c_1 and c_2 using the initial conditions $x(0) = 0$, $x'(0) = 1$ you will get the equations

$$\begin{aligned} c_1 + c_2 &= 0, \\ -2c_1 - 2c_2 &= 1. \end{aligned}$$

These equations are impossible to solve! The moral of the story is that you must have a correct general solution to ensure that you can always solve your IVP.

One more quick example.

Example 1.2.3. Consider the second-order differential equation

$$x'' - x = 0.$$

Suppose we know that the general solution to this differential equation is

$$x(t) = c_1 \exp(t) + c_2 \exp(-t) = c_1 e^t + c_2 e^{-t}$$

for any constants c_1, c_2. (Again, you can check that it is a solution.)

The solution to the IVP

$$x'' - x = 0, \quad x(0) = 1, \; x'(0) = 0,$$

is $x(t) = \frac{e^t+e^{-t}}{2}$. (You can solve for c_1 and c_2 yourself, as in the examples above.) This particular function is also called a *hyperbolic cosine function*, denoted cosh (pronounced "kosh"),

$$\cosh(t) = \frac{e^t + e^{-t}}{2}.$$

The *hyperbolic sine function*, denoted sinh (pronounced "sinch"), satisfies the IVP

$$x'' - x = 0, \quad x(0) = 0, \; x'(0) = 1,$$

and is given by

$$\sinh(t) = \frac{e^t - e^{-t}}{2}.$$

The hyperbolic trigonometric functions have many properties analogous to the usual trigonometric functions and arise in many areas of applications [H-ivp]. For example, $\cosh(t)$ represents a catenary or hanging cable [C-ivp].

Sage

```
sage: t = var('t')
sage: c1 = var('c1')
sage: c2 = var('c2')
sage: de = lambda x: diff(x,t,t) - x
sage: de(c1*exp(-t) + c2*exp(-t))
0
sage: desolve(de(x)),[x,t])
k1*e^t + k2*e^(-t)
sage: P = plot(e^t/2-e^(-t)/2,0,3)
sage: show(P)
```

You see in Figure 1.4 that the solution tends to infinity as t gets larger.

Exercises.

1. Find the value of the constant C that makes $x = Ce^{3t}$ a solution to the IVP $x' = 3x$, $x(0) = 4$.

2. Verify that $x = (C+t)\cos(t)$ satisfies the differential equation $x' + x\tan(t) - \cos(t) = 0$. Find the value of C so that $x(t)$ satisfies the condition $x(2\pi) = 0$.

3. Determine a value of the constant C so that the given solution of the differential equation satisfies the initial condition.

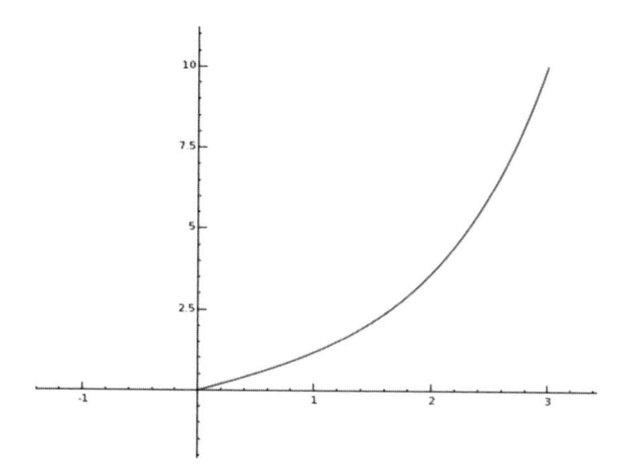

Figure 1.4: Solution to IVP $x'' - x = 0$, $x(0) = 0$, $x'(0) = 1$.

(a) $y = \ln(x + C)$ solves $e^y y' = 1, \quad y(0) = 1$.

(b) $y = Ce^{-x} + x - 1$ solves $y' = x - y, \quad y(0) = 3$.

4. Use Sage to check that the general solution to the falling body problem

$$mv' + kv = mg,$$

is $v(t) = \frac{mg}{k} + ce^{-kt/m}$. If $v(0) = v_0$, you can solve for c in terms of v_0 to get $c = v_0 - \frac{mg}{k}$. Take $m = k = v_0 = 1$, $g = 9.8$ and use Sage to plot $v(t)$ for $0 < t < 1$.

5. Consider a spacecraft falling towards the moon at a speed of 500 m/s (meters per second) at a height of 50 km. Assume the moon's acceleration on the spacecraft is a constant -1.63 m/s^2. When the spacecraft's rockets are engaged, the total acceleration on the rocket (gravity+rocket thrust) is 3 m/s^2. At what height should the rockets be engaged so that the spacecraft lands at zero velocity?

1.3 Existence of solutions to ODEs

When do solutions to an ODE exist? When are they unique? This section gives some necessary conditions for determining existence and uniqueness.

1.3.1 First-order ODEs

We begin by considering the first-order initial value problem

$$x'(t) = f(t, x(t)), \quad x(a) = c. \tag{1.2}$$

What conditions on f (and a and c) guarantee that a solution $x = x(t)$ exists? If it exists, what (further) conditions guarantee that $x = x(t)$ is unique?

The following result addresses the first question.

Theorem 1.3.1. *(Peano's existence theorem [P-intro])* Suppose f is bounded and continuous in x and t. Then, for some value $\epsilon > 0$, there exists a solution $x = x(t)$ to the initial value problem (1.2) within the range $[a - \epsilon, a + \epsilon]$.

Giuseppe Peano (1858–1932) was an Italian mathematician, who is mostly known for his important work on the logical foundations of mathematics. For example, the common notations for union $\cup$ and intersection $\cap$ initially appeared in his first book dealing with mathematical logic, written while he was teaching at the University of Turin.

Example 1.3.1. Take $f(x,t) = x^{2/3}$. This is continuous and bounded in x and t in $-1 < x < 1$, $t \in \mathbb{R}$. The IVP $x' = f(x,t)$, $x(0) = 0$ has two solutions, $x(t) = 0$ and $x(t) = t^3/27$. Therefore, the IVP does not have a unique solution in this example.

You all know what continuity means but you may not be familiar with the slightly stronger notion of "Lipschitz continuity." This is defined next.

Definition 1.3.1. Let $D \subset \mathbb{R}^2$ be a domain. A function $f : D \to \mathbb{R}$ is called *Lipschitz continuous* if there exists a real constant $K > 0$ such that, for all $x_1, x_2 \in D$,

$$|f(x_1) - f(x_2)| \leq K|x_1 - x_2|.$$

The smallest such K is called the *Lipschitz constant* of the function f on D.

For example,

- the function $f(x) = x^{2/3}$ defined on $[-1, 1]$ is not Lipschitz continuous;
- the function $f(x) = x^2$ defined on $[-3, 7]$ is Lipschitz continuous, with Lipschitz constant $K = 14$;
- the function f defined by $f(x) = x^{3/2}\sin(1/x)$ $(x \neq 0)$ and $f(0) = 0$ restricted to $[0, 1]$ gives an example of a function that is differentiable on a compact set while not being Lipschitz.

Theorem 1.3.2. *(Picard's existence and uniqueness theorem [PL-intro])* Suppose f is bounded, Lipschitz continuous in x, and continuous in t. Then, for some value $\epsilon > 0$, there exists a unique solution $x = x(t)$ to the initial value problem (1.2) within the range $[a - \epsilon, a + \epsilon]$.

Charles Émile Picard (1856–1941) was a leading French mathematician. Picard made his most important contributions in the fields of analysis, function theory, differential equations, and analytic geometry. In 1885 Picard was appointed to the mathematics faculty at the Sorbonne in Paris. Picard was awarded the Poncelet Prize in 1886, the Grand Prix des Sciences Mathématiques in 1888, the Grande Croix de la Légion d'Honneur in 1932, and the Mittag-Leffler Gold Medal in 1937, and was made President of the International Congress of Mathematicians in 1920. He is the author of many books and his collected papers run to four volumes.

The proofs of Peano's theorem or Picard's theorem go well beyond the scope of this book. However, for the curious, a very brief indication of the main ideas will be given in the sketch below. For details, see an advanced text on differential equations.

sketch of the idea of the proof: A simple proof of existence of the solution is obtained by successive approximations. In this context, the method is known as Picard iteration.

Set $x_0(t) = c$ and

$$x_i(t) = c + \int_a^t f(s, x_{i-1}(s))\, ds.$$

It turns out that Lipschitz continuity implies that the mapping T defined by

$$T(y)(t) = c + \int_a^t f(s, y(s))\, ds$$

is a contraction mapping on a certain Banach space. It can then be shown, by using the Banach fixed point theorem, that the sequence of "Picard iterates" x_i is convergent and that the limit is a solution to the problem. The proof of uniqueness uses a result called Grönwall's Lemma. □

Example 1.3.2. Consider the IVP

$$x' = 1 - x, \qquad x(0) = 1,$$

with the constant solution $x(t) = 1$. Computing the Picard iterates by hand is easy: $x_0(t) = 1$, $x_1(t) = 1 + \int_0^t 1 - x_0(s)\, ds = 1$, $x_2(t) = 1 + \int_0^t 1 - x_1(s)\, ds = 1$, and so on. Since each $x_i(t) = 1$, we find the solution

$$x(t) = \lim_{i\to\infty} x_i(t) = \lim_{i\to\infty} 1 = 1.$$

We now try the Picard iteration method in Sage. Consider the IVP

$$x' = 1 - x, \qquad x(0) = 2,$$

which we considered earlier.

Sage

```
sage: var('t, s')
sage: f = lambda t,x: 1-x
sage: a = 0; c = 2
sage: x0 = lambda t: c; x0(t)
2
sage: x1 = lambda t: c + integral(f(s,x0(s)), s, a, t); x1(t)
2 - t
sage: x2 = lambda t: c + integral(f(s,x1(s)), s, a, t); x2(t)
t^2/2 - t + 2
sage: x3 = lambda t: c + integral(f(s,x2(s)), s, a, t); x3(t)
-t^3/6 + t^2/2 - t + 2
sage: x4 = lambda t: c + integral(f(s,x3(s)), s, a, t); x4(t)
t^4/24 - t^3/6 + t^2/2 - t + 2
sage: x5 = lambda t: c + integral(f(s,x4(s)), s, a, t); x5(t)
```

```
-t^5/120 + t^4/24 - t^3/6 + t^2/2 - t + 2
sage: x6 = lambda t: c + integral(f(s,x5(s)), s, a, t); x6(t)
t^6/720 - t^5/120 + t^4/24 - t^3/6 + t^2/2 - t + 2
sage: P1 = plot(x2(t), t, 0, 2, linestyle='--')
sage: P2 = plot(x4(t), t, 0, 2, linestyle='-.')
sage: P3 = plot(x6(t), t, 0, 2, linestyle=':')
sage: P4 = plot(1+exp(-t), t, 0, 2)
sage: (P1+P2+P3+P4).show()
```

From the graph in Figure 1.5 you can see how well these iterates are (or at least appear to be) converging to the true solution $x(t) = 1 + e^{-t}$.

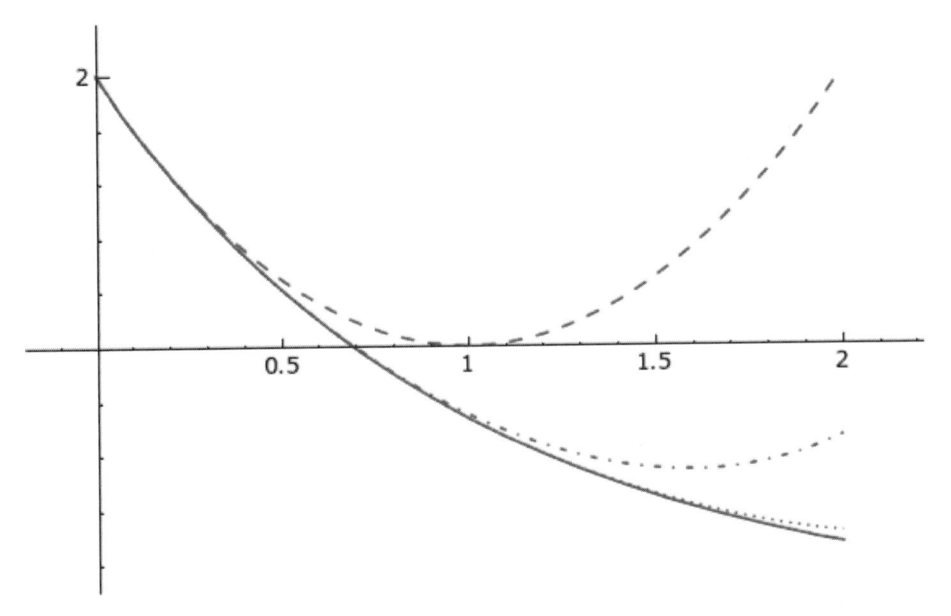

Figure 1.5: Picard iteration for $x' = 1 - x$, $x(0) = 2$.

More generally, here is some Sage code for Picard iteration.

Sage

```
def picard_iteration(f, a, c, N):
    '''
    Computes the N-th Picard iterate for the IVP

        x' = f(t,x), x(a) = c.

    EXAMPLES:
        sage: var('x t s')
        (x, t, s)
        sage: a = 0; c = 2
        sage: f = lambda t,x: 1-x
```

```
        sage: picard_iteration(f, a, c, 0)
        2
         sage: picard_iteration(f, a, c, 1)
        2 - t
        sage: picard_iteration(f, a, c, 2)
        t^2/2 - t + 2
        sage: picard_iteration(f, a, c, 3)
        -t^3/6 + t^2/2 - t + 2

    '''
    if N == 0:
        return c*t**0
    if N == 1:
        x0 = lambda t: c + integral(f(s,c*s**0), s, a, t)
        return expand(x0(t))
    for i in range(N):
        x_old = lambda s: picard_iteration(f, a, c, N-1).subs(t=s)
        x0 = lambda t: c + integral(f(s,x_old(s)), s, a, t)
    return expand(x0(t))
```

Exercises.

1. Compute the first three Picard iterates for the IVP

$$x' = 1 + xt, \qquad x(0) = 0$$

   using the initial function $x_0(t) = 0$.

2. Apply the Picard iteration method in Sage to the IVP

$$x' = (t + x)^2, \qquad x(0) = 2,$$

   and find the first three iterates.

1.3.2 Second-order homogeneous ODEs

We begin by considering the second-order[2] initial value problem

$$ax'' + bx' + cx = 0, \quad x(0) = d_0, \quad x'(0) = d_1, \tag{1.3}$$

where a, b, c, d_0, d_1 are constants and $a \neq 0$. What conditions guarantee that a solution $x = x(t)$ exists? If it exists, what (further) conditions guarantee that $x = x(t)$ is unique? It turns out that no conditions are needed—a solution to (1.3) always exists and is unique.

[2] It turns out that the reasoning in the second-order case is very similar to the general reasoning for nth order differential equations. For simplicity of presentation, we restrict our attention to the second-order case.

As we will see later, we can construct distinct explicit solutions, denoted $x_1 = x_1(t)$ and $x_2 = x_2(t)$ and sometimes called *fundamental solutions*, to $ax'' + bx' + cx = 0$. If we let $x = c_1x_1 + c_2x_2$, for any constants c_1 and c_2, then we know that x is also a solution,[3] sometimes called the *general solution* to $ax'' + bx' + cx = 0$. But how do we know there exist c_1 and c_2 for which this general solution also satisfies the initial conditions $x(0) = d_0$ and $x'(0) = d_1$? For this to hold, we need to be able to solve

$$c_1x_1(0) + c_2x_2(0) = d_1, \qquad c_1x_1'(0) + c_2x_2'(0) = d_2$$

for c_1 and c_2. By Cramer's rule,

$$c_1 = \frac{\begin{vmatrix} d_1 & x_2(0) \\ d_2 & x_2'(0) \end{vmatrix}}{\begin{vmatrix} x_1(0) & x_2(0) \\ x_1'(0) & x_2'(0) \end{vmatrix}}, \qquad c_2 = \frac{\begin{vmatrix} x_1(0) & d_1 \\ x_1'(0) & d_2 \end{vmatrix}}{\begin{vmatrix} x_1(0) & x_2(0) \\ x_1'(0) & x_2'(0) \end{vmatrix}}.$$

For this solution to exist, the denominators in these quotients must be nonzero. This denominator is the value of the "Wronskian" [W-linear] at $t = 0$.

Definition 1.3.2. For n functions $f_1, \ldots, f_n$, which are $n-1$ times differentiable on an interval I, the *Wronskian*[4] $W(f_1, \ldots, f_n)$ as a function on I is defined by

$$W(f_1, \ldots, f_n)(x) = \begin{vmatrix} f_1(x) & f_2(x) & \cdots & f_n(x) \\ f_1'(x) & f_2'(x) & \cdots & f_n'(x) \\ \vdots & \vdots & \ddots & \vdots \\ f_1^{(n-1)}(x) & f_2^{(n-1)}(x) & \cdots & f_n^{(n-1)}(x) \end{vmatrix},$$

for $x \in I$.

The matrix constructed by placing the functions in the first row, the first derivative of each function in the second row, and so on through the $(n-1)$st derivative, is a square matrix sometimes called a *fundamental matrix* of the functions. The Wronskian is the determinant of the fundamental matrix.

Theorem 1.3.3. *(Abel's identity)* Consider a homogeneous linear second-order ordinary differential equation

$$\frac{\mathrm{d}^2y}{\mathrm{d}x^2} + p(x)\frac{\mathrm{d}y}{\mathrm{d}x} + q(x)\,y = 0$$

on the real line with a continuous function p. The Wronskian W of two fundamental solutions of the differential equation satisfies the relation

$$W(x) = W(0)\exp\left(-\int_0^x p(s)\,ds\right).$$

[3]This follows from the linearity assumption.

[4]Josef Wronski was a Polish-born French mathematician who worked in many different areas of applied mathematics and mechanical engineering [Wr-linear].

Example 1.3.3. Consider $x'' + 3x' + 2x = 0$. The Sage code below calculates the Wronskian of the two fundamental solutions e^{-t} and e^{-2t} both directly and from Abel's identity (Theorem 1.3.3).

Sage

```
sage: t = var("t")
sage: x = function("x",t)
sage: DE = diff(x,t,t)+3*diff(x,t)+2*x==0
sage: desolve(DE, [x,t])
k1*e^(-t) + k2*e^(-2*t)
sage: Phi = matrix([[e^(-t), e^(-2*t)],[-e^(-t), -2*e^(-2*t)]]); Phi

[      e^(-t)     e^(-2*t)     ]
[     -e^(-t) -2*e^(-2*t)     ]
sage: W = det(Phi); W
-e^(-3*t)
sage: Wt = e^(-integral(3,t)); Wt
e^(-3*t)
sage: W*W(t=0) == Wt
e^(-3*t) == e^(-3*t)
sage: bool(W*W(t=0) == Wt)
True
```

Definition 1.3.3. We say n functions f_1, ..., f_n are *linearly dependent* over the interval I, if there are numbers $a_1, \dots, a_n$ (not all of them zero) such that

$$a_1 f_1(x) + \cdots + a_n f_n(x) = 0,$$

for $x \in I$. If the functions are not linearly dependent then they are called *linearly independent.*

Theorem 1.3.4. If the Wronskian is non-zero at some point in an interval, then the associated functions are linearly independent on the interval.

Example 1.3.4. If $f_1(t) = e^t$ and $f_2(t) = e^{-t}$ then

$$\begin{vmatrix} e^t & e^{-t} \\ e^t & -e^{-t} \end{vmatrix} = -2.$$

Indeed,

Sage

```
sage: var('t')
t
sage: f1 = exp(t); f2 = exp(-t)
sage: wronskian(f1,f2)
-2
```

Therefore, the fundamental solutions $x_1 = e^t$, $x_2 = e^{-t}$ are linearly independent.

Exercises.

1. Using Sage, verify Abel's identity

   (a) in the example $x'' - x = 0$,

   (b) in the example $x'' + 2 * x' + 2x = 0$.

   Determine what the existence and uniqueness theorem (Theorem 1.3.2) guarantees, if anything, about solutions to the following initial value problems. Note: that you do not have to find the solutions. In some cases you may wish to use Theorem 1.1.1, which is a weaker result, but it is easier to verify the assumptions.

2. $dy/dx = \sqrt{xy}$, $y(0) = 1$.

3. $dy/dx = y^{1/3}$, $y(0) = 2$.

4. $dy/dx = y^{1/3}$, $y(2) = 0$.

5. $dy/dx = x\ln(y)$, $y(0) = 1$.

1.4 First-order ODEs: Separable and linear cases

In this section, we discuss two types of differential equations, both of which are relatively easy to solve.

1.4.1 Separable DEs

We know how to solve any ODE of the form

$$y' = f(t),$$

at least in principle—just integrate both sides.[5] For a more general type of ODE, such as

$$y' = f(t, y),$$

this fails. For instance, if $y' = t + y$ then integrating both sides gives $y(t) = \int \frac{dy}{dt}\,dt = \int y'\,dt = \int t + y\,dt = \int t\,dt + \int y(t)\,dt = \frac{t^2}{2} + \int y(t)\,dt$. So, we have only succeeded in writing $y(t)$ in terms of its integral. Not very helpful.

However, there is a class of ODEs where this idea works, with some modification. If the ODE has the form

$$y' = \frac{g(t)}{h(y)}, \tag{1.4}$$

[5]Recall that y' really denotes $\frac{dy}{dt}$, so by the fundamental theorem of calculus, $y = \int \frac{dy}{dt}\,dt = \int y'\,dt = \int f(t)\,dt = F(t) + c$, where F is the "antiderivative" of f and c is a constant of integration.

then it is called *separable*.[6]

Strategy to solve a separable ODE

(1) Write the ODE (1.4) as $\frac{dy}{dt} = \frac{g(t)}{h(y)}$;

(2) "separate" the t's and the y's:

$$h(y)\,dy = g(t)\,dt;$$

(3) integrate both sides:

$$\boxed{\int h(y)\,dy = \int g(t)\,dt \; + \; C.} \tag{1.5}$$

The "$+C$" is added to emphasize that a constant of integration must be included in your answer (but only on one side of the equation).

The answer obtained in this manner is called an "implicit solution" of (1.4) since it expresses y *implicitly* as a function of t.

Why does this work? It is easiest to understand by working backward from the formula (1.5). Recall that one form of the fundamental theorem of calculus is $\frac{d}{dy}\int h(y)dy = h(y)$. If we think of y as being a function of t and take the t-derivative, we can use the chain rule to get

$$g(t) = \frac{d}{dt}\int g(t)dt = \frac{d}{dt}\int h(y)dy = \left(\frac{d}{dy}\int h(y)dy\right)\frac{dy}{dt} = h(y)\frac{dy}{dt}.$$

So if we differentiate both sides of equation (1.5) with respect to t, we recover the original differential equation.

Example 1.4.1. Are the following ODEs separable? If so, solve them.

(a) $(t^2 + y^2)y' = -2ty$.

(b) $y' = -x/y$, $y(0) = -1$.

(c) $T' = k \cdot (T - T_{room})$, where $k < 0$ and T_{room} are constants.

(d) $ax' + bx = c$, where $a \neq 0$, $b \neq 0$, and c are constants.

(e) $ax' + bx = c$, where $a \neq 0$, b, are constants and $c = c(t)$ is *not* a constant.

(f) $y' = (y-1)(y+1)$, $y(0) = 2$.

---

[6]In particular, any separable differential equation *must* be a first-order ordinary differential equation.

(g) $y' = y^2 + 1$, $y(0) = 1$.

Solutions.

(a) not separable,

(b) $y\,dy = -x\,dx$, so $y^2/2 = -x^2/2 + c$, so $x^2 + y^2 = 2c$. This is the general solution (note that it does not give y explicitly as a function of x; you will have to solve for y algebraically to get that). The initial conditions say that when $x = 0$, $y = 1$, so $2c = 0^2 + 1^2 = 1$, which gives $c = 1/2$. Therefore, $x^2 + y^2 = 1$, which is a circle. That is not a *function* so cannot be the solution we want. The solution is either $y = \sqrt{1-x^2}$ or $y = -\sqrt{1-x^2}$, but which one? Since $y(0) = -1$ (note the minus sign) it must be $y = -\sqrt{1-x^2}$.

(c) $\frac{dT}{T - T_{room}} = kdt$, so $\log|T - T_{room}| = kt + c$ (some constant c), so $T - T_{room} = Ce^{kt}$ (some constant C), so $T = T(t) = T_{room} + Ce^{kt}$.

(d) $\frac{dx}{dt} = (c - bx)/a = -\frac{b}{a}(x - \frac{c}{b})$, so $\frac{dx}{x - \frac{c}{b}} = -\frac{b}{a}\,dt$, so $\log|x - \frac{c}{b}| = -\frac{b}{a}t + C$, where C is a constant of integration. This is the *implicit* general solution of the differential equation. The *explicit* general solution is $x = \frac{c}{b} + Be^{-\frac{b}{a}t}$, where B is a constant.

The explicit solution is easy to find using Sage:

Sage

```
sage: a = var('a')
sage: b = var('b')
sage: c = var('c')
sage: t = var('t')
sage: x = function('x', t)
sage: de = lambda y: a*diff(y,t) + b*y - c
sage: desolve(de(x),[x,t])
(c*e^(b*t/a)/b + c)*e^(-b*t/a)
```

(e) If $c = c(t)$ is not constant then $ax' + bx = c$ is not separable.

(f) $\frac{dy}{(y-1)(y+1)} = dt$ so $\frac{1}{2}(\log(y-1) - \log(y+1)) = t + C$, where C is a constant of integration. This is the "general (implicit) solution" of the differential equation.

Note: the constant functions $y(t) = 1$ and $y(t) = -1$ are also solutions to this differential equation. These solutions cannot be obtained (in an obvious way) from the general solution.

The integration is easy to do using Sage:

Sage

```
sage: y = var('y')
sage: integral(1/((y-1)*(y+1)),y)
log(y - 1)/2 - (log(y + 1)/2)
```

Now, let's try to get Sage to solve for y in terms of t in $\frac{1}{2}(\log(y-1)-\log(y+1)) = t+C$:

Sage

```
sage: C = var('C')
sage: solve([log(y - 1)/2 - (log(y + 1)/2) == t+C],y)
[log(y + 1) == -2*C + log(y - 1) - 2*t]
```

This is not working. Let's try inputting the problem in a different form:

Sage

```
sage: C = var('C')
sage: solve([log((y - 1)/(y + 1)) == 2*t+2*C],y)
[y == (-e^(2*C + 2*t) - 1)/(e^(2*C + 2*t) - 1)]
```

This is what we want. Now let's assume the initial condition $y(0) = 2$ and solve for C and plot the function.

Sage

```
sage: solny=lambda t:(-e^(2*C+2*t)-1)/(e^(2*C+2*t)-1)
sage: solve([solny(0) == 2],C)
[C == log(-1/sqrt(3)), C == -log(3)/2]
sage: C = -log(3)/2
sage: solny(t)
(-e^(2*t)/3 - 1)/(e^(2*t)/3 - 1)
sage: P = plot(solny(t), 0, 1/2)
sage: show(P)
```

This plot is shown in Figure 1.6. The solution has a singularity at $t = \log(3)/2 = 0.5493\ldots$.

(g) $\frac{dy}{y^2+1} = dt$ so $\arctan(y) = t + C$, where C is a constant of integration. The initial condition $y(0) = 1$ says $\arctan(1) = C$, so $C = \frac{\pi}{4}$. Therefore $y = \tan(t + \frac{\pi}{4})$ is the solution.

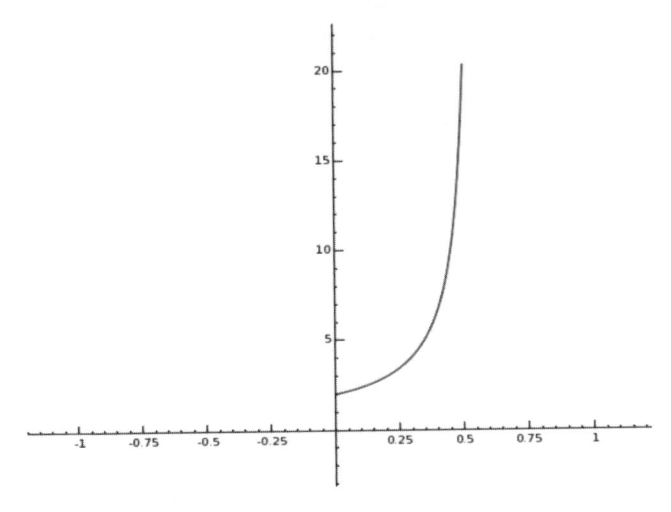

Figure 1.6: Plot of $y' = (y-1)(y+1)$, $y(0) = 2$, for $0 < t < 1/2$.

1.4.2 Autonomous ODEs

A special subclass of separable ODEs is the class of *autonomous* ODEs, which have the form

$$y' = f(y),$$

where f is a given function (i.e., the slope y' depends only on the value of the dependent variable y). The cases (c), (d), (f), and (g) above are examples.

One of the simplest examples of an autonomous ODE is $\frac{dy}{dt} = ky$, where k is a constant. We can divide by y and integrate to get

$$\int \frac{1}{y}\, dy = \log|y| = \int k\, dt = kt + C_1.$$

After exponentiating, we get

$$|y| = e^{kt+C_1} = e^{C_1}e^{kt}.$$

We can drop the absolute value if we allow positive and negative solutions:

$$y = \pm e^{C_1}e^{kt}.$$

Now note that $\pm e^{C_1}$ can be any nonzero number, but in fact $y = 0$ is also a solution to the ODE so we can write the general solution as $y = Ce^{kt}$ for an arbitrary constant C. If k is positive, solutions will grow in magnitude exponentially, and if k is negative solutions will decay to 0 exponentially.

Perhaps the most famous use of this type of ODE is in carbon dating.

Example 1.4.2. Carbon-14 (^{14}C) has a half-life of about 5730 years, meaning that after that time one-half of a given amount will radioactively decay (into stable nitrogen-14).

Prior to the nuclear tests of the 1950s, which raised the level of ^{14}C in the atmosphere, the ratio of ^{14}C to Carbon-12 (^{12}C) in the air, plants, and animals was 10^{-15}. If this ratio is measured in an archeological sample of bone and found to be 10^{-17}, how old is the sample?

Solution. Since a constant fraction of ^{14}C decays per unit time, the amount of ^{14}C satisfies a linear differential equation $y' = ky$ with solution $y = Ce^{kt}$. Since

$$y(5730) = Ce^{k5730} = y(0)/2 = C/2,$$

we can compute $k = -\log(2)/5730 \approx 1.21 \cdot 10^{-5}$. Let t_0 denote the time of death of whatever the bone sample is from. We know that ^{12}C is stable, so if we divide the ratio of ^{14}C to ^{12}C at $t = 0$ into that ratio at $t = t_0$, the ^{12}C amounts will cancel. Therefore, we know that

$$y(t_0)/y(0) = \frac{10^{-17}}{10^{-15}} = 10^{-2} = \frac{Ce^{kt_0}}{C} = e^{kt_0}.$$

This gives $t_0 = \frac{\log 10^2}{k} \approx -38069$ years, so the bone sample is about 38000 years old.

Here is a nonlinear example.

Example 1.4.3. Consider

$$y' = (y-1)(y+1), \qquad y(0) = 1/2.$$

Here is one way to solve this using Sage:

```
Sage
sage: t = var('t')
sage: x = function('x', t)
sage: de = lambda y: diff(y,t) == y^2 - 1
sage: soln = desolve(de(x),[x,t]); soln
1/2*log(x(t) - 1) - 1/2*log(x(t) + 1) == c + t
sage:   # needs an abs. value ...
sage: c,xt = var("c,xt")
sage: solnxt = (1/2)*log(abs(xt - 1)) - (1/2)*log(abs(xt + 1))
                        == c + t
sage: solve(solnxt.subs(t=0, xt=1/2),c)
[c == -1/2*log(3/2) - 1/2*log(2)]
sage: c0 = solve(solnxt.subs(t=0, xt=1/2),c)[0].rhs(); c0
-1/2*log(3/2) - 1/2*log(2)
sage: soln0 = solnxt.subs(c=c0); soln0
1/2*log(abs(xt - 1)) - 1/2*log(abs(xt + 1))
     ==  t - 1/2*log(3/2) -1/2*log(2)
sage: implicit_plot(soln0,(t,-1/2,1/2),(xt,0,0.9))
```

Sage cannot solve this (implicit) solution for $x(t)$, though we're sure you can do it by hand if you want. The (implicit) plot is given in Figure 1.7.

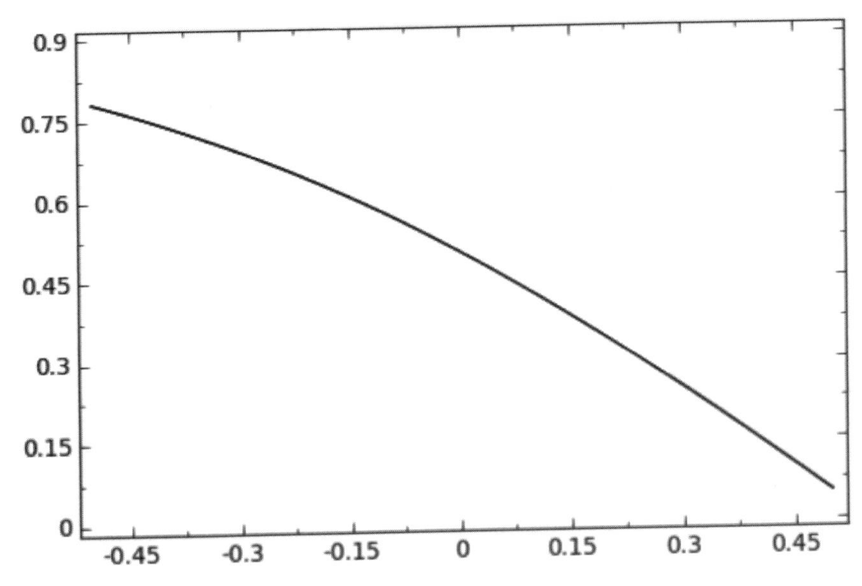

Figure 1.7: Plot of $y' = (y-1)(y+1)$, $y(0) = 1/2$, for $-1/2 < t < 1/2$.

A more complicated example is

$$y' = y(y-1)(y-2).$$

This has constant solutions $y(t) = 0$, $y(t) = 1$, and $y(t) = 2$. (Check this.) Several nonconstant solutions are plotted in Figure 1.8.

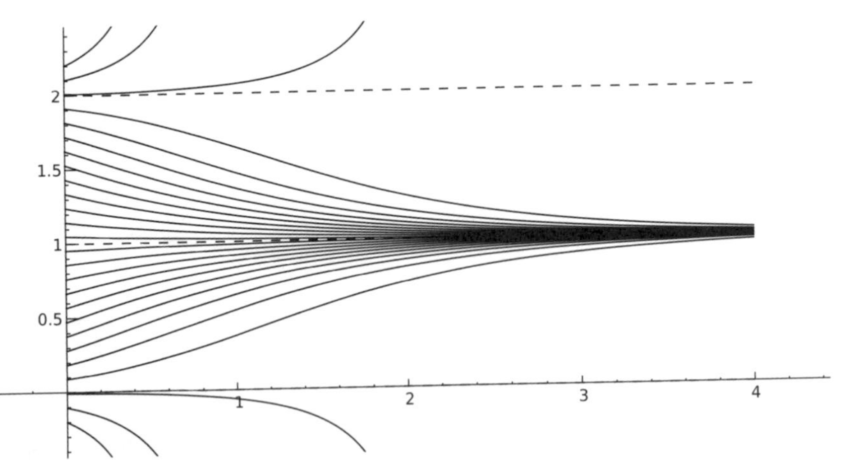

Figure 1.8: Plot of $y' = y(y-1)(y-2)$, $y(0) = y_0$, for $0 < t < 4$, and various values of y_0.

Exercises

1. Find the general solution to $y' = y(y-1)(y-2)$ either by hand or using Sage.

2. Find the general solution to $y' = y^2 - 4$ either by hand or using `Sage`.

3. $x' + x = 1$, $x(0) = 0$.

1.4.3 Substitution methods

Sometimes it is possible to solve differential equations by substitution, just as some integrals can be simplified by substitution.

Example 1.4.4. We will solve the ODE $y' = (y+x+3)^2$ with the substitution $v = y+x+3$. Differentiating the substitution gives the relation $v' = y'+1$, so the ODE can be rewritten as $v'-1 = v^2$, or $v' = v^2+1$. This transformed ODE is separable, so we divide by v^2+1 and integrate to get $\arctan(v) = x + C$. Taking the tangent of both sides gives $v = \tan(x+C)$, and now we substitute back to get $y = \tan(x+C) - x - 3$.

There are a variety of very specialized substitution methods which we will not present in this text. However one class of differential equations is common enough to warrant coverage here: *homogeneous* ODEs. Unfortunately, for historical reasons the word homogeneous has come to mean two different things. In the present context, we define a homogeneous first-order ODE to be one that can be written as $y' = f(y/x)$. For example, the following ODE is homogeneous in this sense:

$$\frac{dy}{dx} = x/y.$$

A first-order homogeneous ODE can be simplified by using the substitution $v = y/x$, or equivalently $y = vx$. Differentiating that relationship gives us $v + xv' = y'$.

Example 1.4.5. We will solve the ODE $y' = y/x + 2$ with the substitution $v = y/x$.
As noted above, with this substitution $y' = v+xv'$ so the original ODE becomes $v+xv' = v+2$. This simplifies to $v' = 2/x$ which can be directly integrated to get $v = 2\log(x) + C$ (in more difficult examples we would get a separable ODE after the substitution). Using the substitution once again we obtain $y/x = 2\log(x) + C$ so the general solution is $y = 2x\log(x) + Cx$.

1.4.4 Linear first-order ODEs

The bottom line is that we want to solve any problem of the form

$$x' + p(t)x = q(t), \tag{1.6}$$

where $p(t)$ and $q(t)$ are given functions (which, let's assume, aren't "too horrible"). Every first-order linear ODE can be written in this form. Examples of differential equations which have this form are falling body problems, Newton's law of cooling problems, mixing problems, certain simple circuit problems, and so on.

There are two approaches:

- "the formula";

- the method of integrating factors.

Both lead to the exact same solution.
The formula. The general solution to (1.6) is

$$x = \frac{\int e^{\int p(t)\,dt} q(t)\,dt + C}{e^{\int p(t)\,dt}}, \tag{1.7}$$

where C is a constant. The factor $e^{\int p(t)\,dt}$ is called the *integrating factor* and is often denoted by μ. This formula was apparently first discovered by Johann Bernoulli [F-1st].

Example 1.4.6. Solve

$$xy' + y = e^x.$$

We rewrite this as $y' + \frac{1}{x}y = \frac{e^x}{x}$. Now compute $\mu = e^{\int \frac{1}{x}\,dx} = e^{\log(x)} = x$, so the formula gives

$$y = \frac{\int x\frac{e^x}{x}\,dx + C}{x} = \frac{\int e^x\,dx + C}{x} = \frac{e^x + C}{x}.$$

Here is one way to do this using Sage:

Sage

```
sage: t = var('t')
sage: x = function('x', t)
sage: de = lambda y: diff(y,t) + (1/t)*y - exp(t)/t
sage: desolve(de(x),[x,t])
(c + e^t)/t
```

Integrating factor method. Let $\mu = e^{\int p(t)\,dt}$. Multiply both sides of (1.6) by μ:

$$\mu x' + p(t)\mu x = \mu q(t).$$

The product rule implies that

$$(\mu x)' = \mu x' + p(t)\mu x = \mu q(t).$$

(In response to a question you are probably thinking now: No, this is not obvious. This is Bernoulli's very clever idea.) Now just integrate both sides. By the fundamental theorem of calculus,

$$\mu x = \int (\mu x)'\,dt = \int \mu q(t)\,dt.$$

Dividing both side by μ gives (1.7).

The integrating factor can be useful in organizing a computation. There are four main steps to solving a first-order linear equation; the integral in the third step is usually the worst part:

1. Put the equation in the standard form $x' + p(t)x = q(t)$.
2. Compute $\mu = e^{\int p(t)\,dt}$.
3. Compute $\displaystyle\int \mu q(t)\,dt$.
4. Combine into the general solution $x = \dfrac{C}{\mu} + \dfrac{1}{\mu}\displaystyle\int \mu q(t)\,dt$.

When solving initial value problems with $x(t_0) = x_0$ it can be convenient to do each of the integrations above from t_0 to t, and then the constant C will be x_0. In other words, if $x(t_0) = x_0$ and $x' + p(t)x = q(t)$ then let $\mu(t) = e^{\int_{t_0}^t p(u)\,du}$ and

$$x = \frac{\int_{t_0}^t \mu(u)q(u)\,du + x_0}{\mu(t)}.$$

Example 1.4.7. Solve the initial value problem

$$t\log(t)x' = -x + t(\log(t))^2, \quad x(e) = 1.$$

First we add x to each side of this equation and divide by $t\log(t)$ to get it into standard form:

$$x' + \frac{1}{t\log(t)}x = \log(t).$$

So $p(t) = \dfrac{1}{t\log(t)}$ and $q(t) = \log(t)$.

Next we compute the integrating factor. Since we are solving an initial value problem we will integrate from e to t:

$$\mu = e^{\int_e^t \frac{1}{u\log(u)}\,du} = e^{\log(\log(t)) - \log(\log(e))} = \log(t).$$

Now we need

$$\int_e^t \mu(u)q(u)\,du = \int_e^t (\log(u))^2\,du = t\,(\log t)^2 - 2t\,(\log t) + 2t - e.$$

Finally we get

$$x(t) = \frac{\int_{t_0}^t \mu(u)q(u)\,du + x_0}{\mu(t)} = t\log(t) - 2t + \frac{2t - e + 1}{\log(t)}.$$

Exercises

Find the general solution to the following separable differential equations:

1. $x' = 2xt$.

2. $x' - x\sin(t) = 0$.

3. $(1+t)x' = 2x$.

4. $x' - xt - x = 1 + t$.

5. Find the solution $y(x)$ to the following initial value problem: $y' = 2ye^x$, $y(0) = 2e^2$.

6. Find the solution $y(x)$ to the following initial value problem: $y' = x^3(y^2+1)$, $y(0) = 1$.

7. Use the substitution $v = y/x$ to solve the IVP $y' = \frac{2xy}{x^2-y^2}$, $y(0) = 2$.

   Solve the following linear equations. Find the general solution if no initial condition is given.

8. $x' + 4x = 2te^{-4t}$.

9. $tx' + 2x = 2t$, $x(1) = \frac{1}{2}$.

10. $dy/dx + y\tan(x) = \sin(x)$.

11. $xy' = y + 2x$, $y(1) = 2$.

12. $y' + 4y = 2xe^{-4x}$, $y(0) = 0$.

13. $y' = \cos(x) - y\cos(x)$, $y(0) = 1$.

14. In carbon-dating organic material it is assumed that the amount of carbon-14 (^{14}C) decays exponentially ($\frac{d}{dt}(^{14}\text{C}) = -k \cdot {}^{14}\text{C}$) with rate constant of $k \approx 0.0001216$ where t is measured in years. Suppose an archeological bone sample contains 1/7 as much carbon-14 as is in a recent sample (but which dates from before 1950). How old is the bone?

15. The function e^{-t^2} does not have an antiderivative in terms of elementary functions, but this antiderivative is important in probability. Define the *error function* by

$$\text{erf}(t) = \frac{2}{\sqrt{\pi}} \int_0^t e^{-u^2}\,du.$$

   Find the solution of $x' - 2xt = 1$ in terms of $\text{erf}(t)$.

16. (a) Use the command `desolve` in `Sage` to solve

$$tx' + 2x = e^t/t.$$

   (b) Use `Sage` to plot the solution to $y' = y^2 - 1$, $y(0) = 2$.

1.5 Isoclines and direction fields

Recall from vector calculus the notion of a two-dimensional vector field: $\vec{F}(x,y) = (g(x,y), h(x,y))$. To plot $\vec{F}$, you simply draw the vector $\vec{F}(x,y)$ at each point (x,y).

The idea of the *direction field* (or *slope field*) associated with the first-order ODE

$$y' = f(x,y), \quad y(a) = c \tag{1.8}$$

is similar. At each point (x,y) you plot a small vector having slope $f(x,y)$. For example, the vector field plot of $\vec{F}(x,y) = (1, f(x,y))$ or, better yet, $\vec{F}(x,y) = (1, f(x,y))/\sqrt{1+f(x,y)^2}$ (which is a unit vector) would work for the slope field.

How would you draw such a direction field plot *by hand*? You could compute the value of $f(x,y)$ for lots and lots of points (x,y) and then plot a tiny arrow of slope $f(x,y)$ at each of these points. Unfortunately, this would be impossible for a person to do on a practical level if the number of points was large.

For this reason, the notion of the "isoclines" of the ODE is very useful. An *isocline* of (1.8) is a level curve of the function $z = f(x,y)$:

$$\{(x,y) \mid f(x,y) = m\},$$

where the given constant m is called the *slope* of the isocline. In terms of the ODE, this curve represents the collection of all points (x,y) at which the solution has slope m. In terms of the direction field of the ODE, it represents the collection of points where the vectors have slope m. This means that once you have draw a single isocline, you can sketch ten or more tiny vectors describing your direction field. Very useful indeed! This idea is recoded below more algorithmically.

How to draw the direction field of (1.8) by hand

- Draw several isoclines, making sure to include one which contains the point (a,c). (You may want to draw these in pencil.)
- On each isocline, draw hatch marks or arrows along the line each having slope m.

This is a crude direction field plot. The plot of arrows forms your direction field. The isoclines, having served their purpose, can now safely be ignored.

Example 1.5.1. The direction field, with three isoclines, for

$$y' = 5x + y - 5, \quad y(0) = 1,$$

is given by the graph in Figure 1.9.

The isoclines are the curves (coincidentally, lines) of the form $5x + y - 5 = m$. These are lines of slope -5, not to be confused with the fact that the plot represents an isocline of slope m.

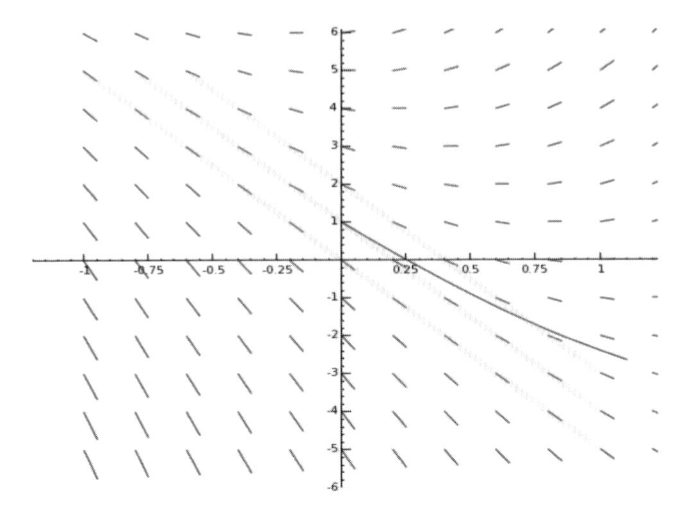

Figure 1.9: Plot of $y' = 5x + y - 5$, $y(0) = 1$, for $-1 < x < 1$.

The above example can be solved explicitly. (Indeed, $y = -5x + e^x$ solves $y' = 5x + y - 5$, $y(0) = 1$.) In the next example, such an explicit solution is not possible. Therefore, a numerical approximation plays a more important role.

Example 1.5.2. The direction field, with three isoclines, for

$$y' = x^2 + y^2, \quad y(0) = 3/2,$$

is given in Figure 1.10.

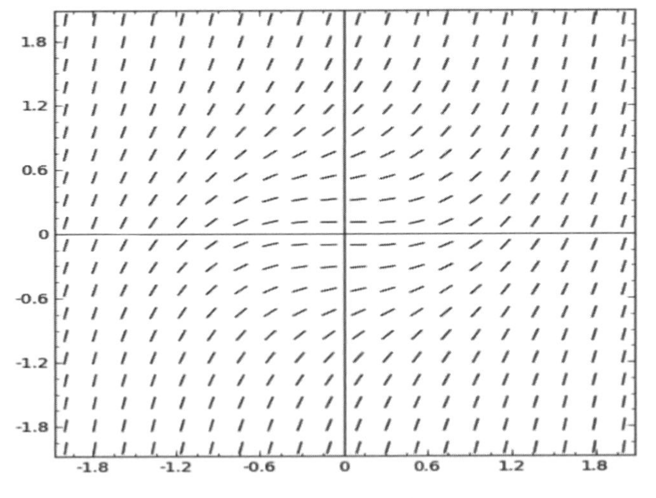

Figure 1.10: Direction field of $y' = x^2 + y^2$, for $-2 < x < 2$.

The isoclines are the concentric circles $x^2 + y^2 = m$.

The plot in Figure 1.10 was obtaining using the Sage code below.

Sage

```
sage: x,y = var("x,y")
sage: f(x,y) = x^2 + y^2
sage: plot_slope_field(f(x,y), (x,-2,2),(y,-2,2)).show(aspect_ratio=1)
```

There is also a way to "manually draw" these direction fields using Sage.

Sage

```
sage: pts = [(-2+i/5,-2+j/5) for i in range(20) \
                    for j in range(20)] # square [-2,2]x[-2,2]
sage: f = lambda p:p[0]^2+p[1]^2 # x = p[0] and y = p[1]
sage: arrows = [arrow(p, (p[0]+0.02,p[1]+(0.02)*f(p)), \
                  width=1/100, rgbcolor=(0,0,1)) for p in pts]
sage: show(sum(arrows))
```

This gives the plot in Figure 1.11.

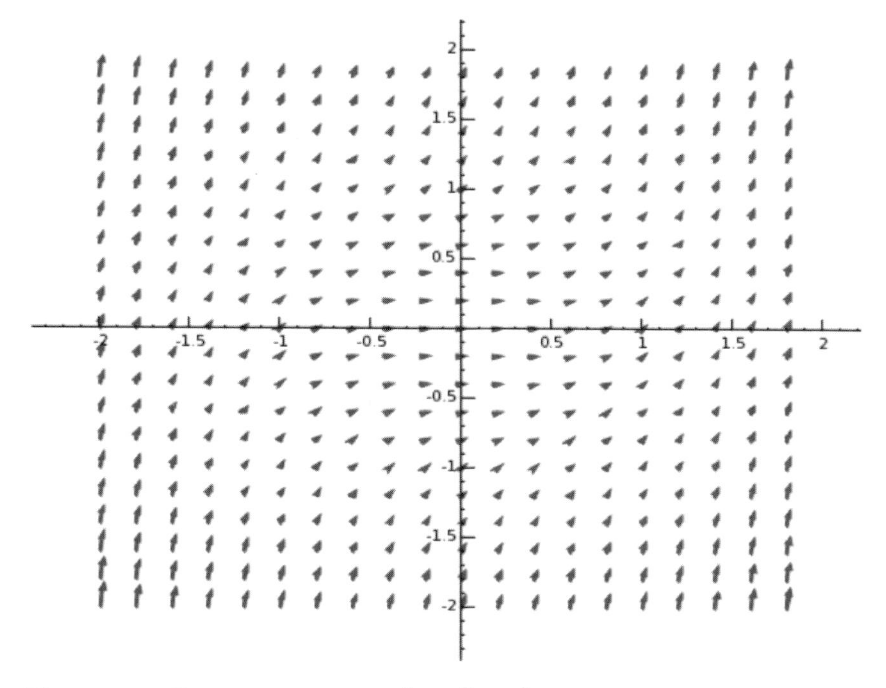

Figure 1.11: Direction field for $y' = x^2 + y^2$, $y(0) = 3/2$, for $-2 < x < 2$.

The plot in Figure 1.12 was obtaining using the Sage code below (the Euler method command is explained in the next section).

Sage

```
sage: x,y = var("x,y")
sage: f(x,y) = x^2 + y^2
sage: P1 = plot_slope_field(f(x,y), (x,-2,2),(y,-2,2))
sage: from sage.calculus.desolvers import eulers_method
sage: x,y = PolynomialRing(RR,2,"xy").gens()
sage: pts = eulers_method(x^2+y^2,-1,0,1/10,1,algorithm="none")
sage: P2 = line(pts, xmin=-2,xmax=2,ymin=-2,ymax=2, thickness=2)
sage: show(P1+P2, aspect_ratio=1, dpi=300)
```

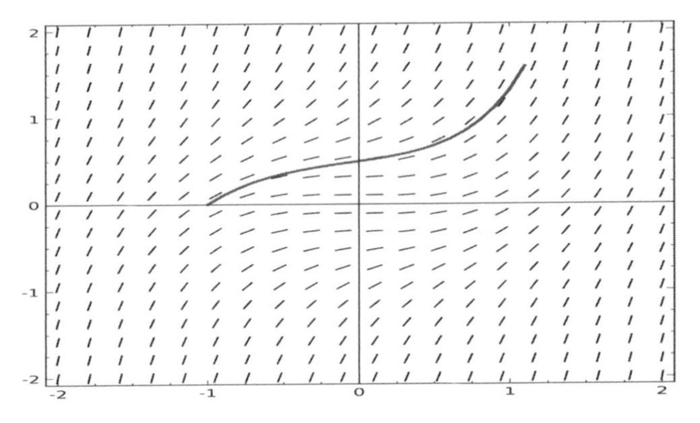

Figure 1.12: Direction field for $y' = x^2 + y^2$, $y(-1) = 0$, for $-2 < x < 2$.

Exercises

1. Match the solution curves in Figure 1.13 to the ODEs below.

   I. $y' = y^2 - 1$.

   II. $y' = \frac{y}{t^2} - 1$.

   III. $y' = \sin(t)\sin(y)$.

   IV. $y' = \sin(ty)$.

   V. $y' = 2t + y$.

   VI. $y' = \sin(3t)$.

2. Use Sage to plot the direction field for $y' = x^2 - y^2$.

3. Using the direction field in Figure 1.10 , sketch the solution to $y' = x^2+y^2$, $y(0) = 3/2$, for $-2 < x < 2$.

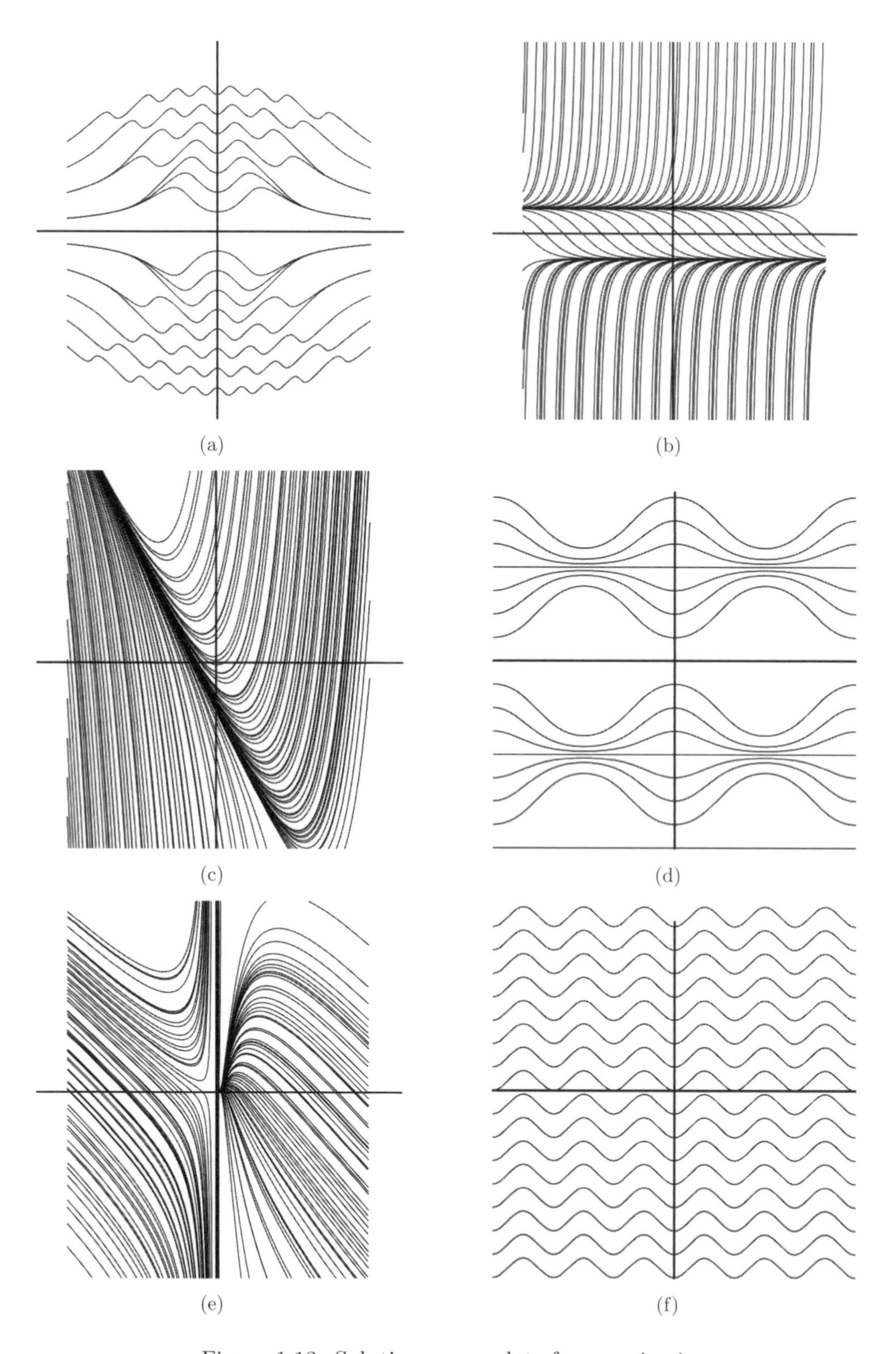

Figure 1.13: Solution curve plots for exercise 1

1.6 Numerical solutions: Euler's and improved Euler's method

> Read Euler: he is our master in everything.
>
> —*Pierre Simon de Laplace*

Leonhard Euler was a Swiss mathematician who made significant contributions to a wide range of mathematics and physics including calculus and celestial mechanics (see [Eu1-num] and [Eu2-num] for further details).

1.6.1 Euler's method

The goal is to find an approximate solution to the problem

$$y' = f(x, y), \quad y(a) = c, \tag{1.9}$$

where $f(x, y)$ is some given function. We shall try to approximate the value of the solution at $x = b$, where $b > a$ is given. Sometimes such a method is called "numerical integration" of (1.9).

Note: the first order differential equation must be in the form (1.9) or the method described below does not work. A version of Euler's method for systems of first-order and higher-order differential equations will also be described below.

Geometric idea

The basic idea can be easily expressed in geometric terms. We know that the solution, whatever it is, must go through the point (a, c) and we know, at that point, its slope is $m = f(a, c)$. Using the point-slope form of a line, we conclude that the tangent line to the solution curve at (a, c) is (in (x, y)-coordinates, not to be confused with the dependent variable y and independent variable x of the differential equation)

$$y = c + (x - a)f(a, c).$$

(See the plot given in Figure 1.14.) In particular, if $h > 0$ is a given small number (called the *increment* or *step size*) then, taking $x = a + h$, the tangent-line approximation from differential calculus gives us

$$y(a + h) \cong c + h \cdot f(a, c).$$

Now we know the solution passes through a point which is "nearly" equal to $(a + h, c + h \cdot f(a, c))$. We now repeat this tangent-line approximation with (a, c) replaced by $(a + h, c + h \cdot f(a, c))$. Keep repeating this number crunching at $x = a$, $x = a + h$, $x = a + 2h$, $\ldots$, until you get to $x = b$.

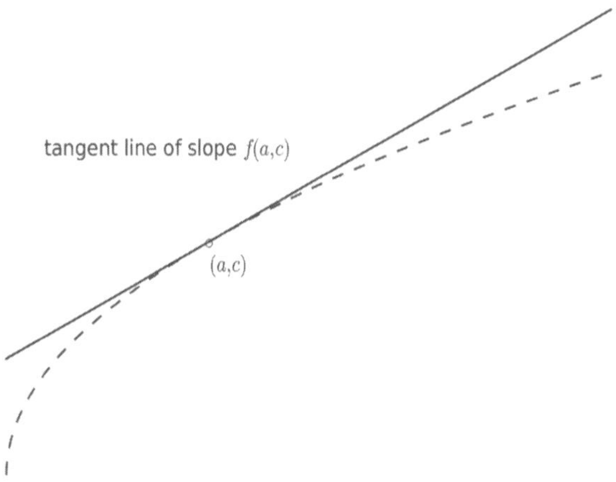

Figure 1.14: Tangent line (solid line) of a solution (dotted curve) to a 1st order DE, (1.9).

Algebraic idea

The basic idea can also be explained "algebraically." Recall from the definition of the derivative in calculus 1 that

$$y'(x) \cong \frac{y(x+h) - y(x)}{h},$$

where $h > 0$ is a given and small. This and the differential equation together give $f(x, y(x)) \cong \frac{y(x+h)-y(x)}{h}$. Now solve for $y(x + h)$:

$$y(x + h) \cong y(x) + h \cdot f(x, y(x)).$$

If we call $h \cdot f(x, y(x))$ the "correction term" (for lack of anything better), call $y(x)$ the "old value of y," and call $y(x + h)$ the "new value of y," then this approximation can be re-expressed

$$y_{new} = y_{old} + h \cdot f(x, y_{old}).$$

Tabular idea

Let $n > 0$ be an integer, the number of steps we will take from the initial condition to the final approximation. This is related to the increment h by

$$h = \frac{b - a}{n}.$$

This can be expressed most simply using a table.

	x	y	$hf(x,y)$
step 0	a	c	$hf(a,c)$
step 1	$a+h$	$c+hf(a,c)$	$\vdots$
step 2	$a+2h$	$\vdots$	
$\vdots$			
step n	b	???	xxx

The goal is to fill out all the blanks of the table but the xxx entry and find the ??? entry, which is the *Euler's method approximation for* $y(b)$.

Example 1.6.1. Use Euler's method with $h = 1/2$ to approximate $y(1)$, where

$$y' - y = 5x - 5, \quad y(0) = 1.$$

Putting the differential equation into the form (1.9), we see that here $f(x, y) = 5x + y - 5$, $a = 0$, $c = 1$.

x	y	$hf(x,y) = \frac{1}{2}(5x+y-5)$
0	1	-2
1/2	$1 + (-2) = -1$	$-7/4$
1	$-1 + (-7/4) = -11/4$	

Thus $y(1) \cong -\frac{11}{4} = -2.75$. This is the final answer.

Aside: For your information, $y = e^x - 5x$ solves the differential equation and $y(1) = e - 5 = -2.28\ldots$. You see from this that the numerical approxmation is not too far away from the actual value of the solution. Not bad, for only two steps!

Here is one way to do this using Sage:

Sage

```
sage: x,y=PolynomialRing(QQ,2,"xy").gens()
sage: eulers_method(5*x+y-5,1,1,1/3,2)
         x                    y                 h*f(x,y)
         1                    1                      1/3
       4/3                  4/3                        1
       5/3                  7/3                     17/9
         2                 38/9                    83/27
sage: eulers_method(5*x+y-5,0,1,1/2,1,method="none")
[[0, 1], [1/2, -1], [1, -11/4], [3/2, -33/8]]
sage: pts = eulers_method(5*x+y-5,0,1,1/2,1,method="none")
sage: P = list_plot(pts)
sage: show(P)
sage: P = line(pts)
sage: show(P)
sage: P1 = list_plot(pts)
sage: P2 = line(pts)
```

```
sage: show(P1+P2)
```

The plot is given in Figure 1.15.

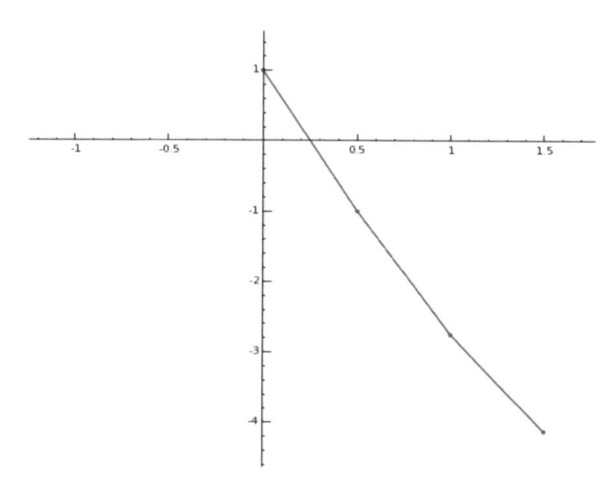

Figure 1.15: Euler's method with $h = 1/2$ for $y' - y = 5x - 5$, $y(0) = 1$.

1.6.2 Improved Euler's method

Geometric idea

The basic idea can be easily expressed in geometric terms. As in Euler's method, we know that the solution must go through the point (a, c) and we know that its slope there is

$$m = f(a, c).$$

If we went out one step using the tangent-line approximation to the solution curve, the approximate slope to the tangent-line at $x = a + h, y = c + h \cdot f(a, c)$ would be

$$m' = f(a + h, c + h \cdot f(a, c)).$$

The idea is that instead of using $m = f(a, c)$ as the slope of the line to get our first approximation, we use $\frac{m+m'}{2}$. The "improved" tangent-line approximation at (a, c) is

$$y(a + h) \cong c + h \cdot \frac{m + m'}{2} = c + h \cdot \frac{f(a, c) + f(a + h, c + h \cdot f(a, c))}{2}.$$

(This turns out to be a better approximation than the tangent-line approximation $y(a+h) \cong c+h \cdot f(a, c)$ used in Euler's method.) Now we know the solution passes through a point which is "nearly" equal to $(a + h, c + h \cdot \frac{m+m'}{2})$. We now repeat this tangent-line approximation with (a, c) replaced by $(a + h, c + h \cdot f(a, c))$. Keep repeating this number crunching at $x = a$, $x = a + h$, $x = a + 2h$, $\ldots$, until you get to $x = b$.

Tabular idea

The integer step size $n > 0$ is related to the increment by

$$h = \frac{b-a}{n},$$

as before.

The improved Euler's method can be expressed most simply using a table.

x	y	$\frac{h}{2}(f(x,y) + f(x+h, y+h\cdot f(x,y)))$
a	c	$\frac{h}{2}(f(a,c) + f(a+h, c+h\cdot f(a,c)))$
$a+h$	$c + \frac{h}{2}(f(a,c) + f(a+h, c+h\cdot f(a,c)))$	$\vdots$
$a+2h$	$\vdots$	
$\vdots$		
b	???	xxx

The goal is to fill out all the blanks of the table but the xxx entry and find the ??? entry, which is the *improved Euler's method approximation for $y(b)$*.

Example 1.6.2. Use the improved Euler's method with $h = 1/2$ to approximate $y(1)$, where

$$y' - y = 5x - 5, \quad y(0) = 1.$$

Putting the differential equation into the form (1.9), we see that here $f(x,y) = 5x + y - 5$, $a = 0$, $c = 1$. We first compute the "correction term":

$$\begin{aligned}
h\frac{f(x,y)+f(x+h,y+h\cdot f(x,y))}{2} &= \tfrac{1}{4}(5x + y - 5 + 5(x+h) + (y + h\cdot f(x,y)) - 5) \\
&= \tfrac{1}{4}(5x + y - 5 + 5(x+h) + (y + h\cdot(5x+y-5) - 5) \\
&= ((1+\tfrac{h}{2})5x + (1+\tfrac{h}{2})y - \tfrac{5}{2})/2 \\
&= (25x/4 + 5y/4 - 5)/2.
\end{aligned}$$

and then use this in two steps of improved Euler's method:

x	y	$\frac{25x+5y-20}{8}$
0	1	$-15/8$
1/2	$1 + (-15/8) = -7/8$	$-95/64$
1	$-7/8 + (-95/64) = -151/64$	

This gives the approximation $y(1) \cong -\frac{151}{64} = -2.35\ldots$. This is the final answer.

Aside: For your information, this is closer to the exact value $y(1) = e - 5 = -2.28\ldots$ than the "usual" Euler's method approximation of -2.75 that we obtained above.

Here is one way to do this using Sage.

Sage

```
def improved_eulers_method(f,x0,y0,h,x1):
    print "%10s %30s %50s"%("x","y","(h/2)*(f(x,y)+f(x+h,y+h*f(x,y))")
    n=((RealField(max(6,RR.precision()))('1.0'))*(x1-x0)/h).integer_part()
    x00=x0; y00=y0
    for i in range(n+ZZ(1)):
        s1 = f(x00,y00)
        s2= f(x00+h,y00+h*f(x00,y00))
        corr = h*(ZZ(1)/ZZ(2))*(s1+s2)
        print "%10r %30r %30r"%(x00,y00,corr)
        y00 = y00+h*(ZZ(1)/ZZ(2))*(f(x00,y00)+f(x00+h,y00+h*f(x00,y00)))
        x00=x00+h
```

This implements the improved Euler's method for finding the "numerical solution" of the first-order ODE $y' = f(x, y)$, $y(a) = c$. The x-column of the table increments from $x = 0$ to $x = 1$ by h. In the y-column, the new y-value equals the old y-value plus the corresponding entry in the last column.

Sage

```
  sage: RR = RealField(sci_not=0, prec=4, rnd='RNDU')
  sage: x,y = PolynomialRing(RR,'x,y',2).gens()
  sage: improved_eulers_method(5*x+y-5,0,1,1/2,1)
     x                                    y                  (h/2)*(f(x,y)+f(x+h,y+h*f(x,y))
     0                                    1                                    -1.87
   1/2                               -0.875                                    -1.37
     1                                -2.25                                   -0.687
  sage: x,y=PolynomialRing(QQ,'x,y',2).gens()
  sage: improved_eulers_method(5*x+y-5,0,1,1/2,1)
     x                                    y                  (h/2)*(f(x,y)+f(x+h,y+h*f(x,y))
     0                                    1                                    -15/8
   1/2                                 -7/8                                   -95/64
     1                              -151/64                                 -435/512
```

This yields the approximation $y(1) \approx -0.875$.

Example 1.6.3. Use both Euler's method and the improved Euler's method with $h = 1/2$ to approximate $y(2)$, where y solves

$$y' = xy, \quad y(1) = 1. \tag{1.10}$$

x	y	$hf(x,y) = xy/2$	y	$h(f(x,y) + f(x+h, y+hf(x,y))/2)$ $= 9xy/16 + y/8 + x^2y/8$
1	1	1/2	1	13/16
3/2	3/2	9/8	29/16	145/64
2	21/8	xxx	261/64	xxx

These give, for Euler's method $y(2) \approx 21/8 = 2.625$, and for improved Euler's method $y(2) \approx 261/64 = 4.078\dots$. The exact solution $y(2) = e^{3/2} = 4.4816\dots$, since the solution to (1.10) is $y = e^{x^2/2-1/2}$ (which the reader is asked to verify for him/herself). Clearly, the improved Euler's method gives the better approximation in this case.

1.6.3 Euler's method for systems and higher-order DEs

We sketch the idea only in some simple cases. Consider the differential equation

$$y'' + p(x)y' + q(x)y = f(x), \quad y(a) = e_1, \; y'(a) = e_2,$$

and the system

$$\begin{aligned} y_1' &= f_1(x, y_1, y_2), \quad y_1(a) = c_1, \\ y_2' &= f_2(x, y_1, y_2), \quad y_2(a) = c_2. \end{aligned}$$

We can treat both cases after first rewriting the second-order differential equation as a system: create new variables $y_1 = y$ and let $y_2 = y'$. It is easy to see that

$$\begin{aligned} y_1' &= y_2, & y_1(a) &= e_1, \\ y_2' &= f(x) - q(x)y_1 - p(x)y_2, & y_2(a) &= e_2. \end{aligned}$$

Tabular idea

Let $n > 0$ be an integer, which we call the *step size*. This is related to the increment by

$$h = \frac{b-a}{n}.$$

This can be expressed most simply using a table.

x	y_1	$hf_1(x, y_1, y_2)$	y_2	$hf_2(x, y_1, y_2)$
a	e_1	$hf_1(a, e_1, e_2)$	e_2	$hf_2(a, e_1, e_2)$
$a+h$	$e_1 + hf_1(a, e_1, e_2)$	$\vdots$	$e_2 + hf_2(a, e_1, e_2)$	$\vdots$
$a+2h$	$\vdots$			
$\vdots$				
b	???	xxx	xxx	xxx

The goal is to fill out all the blanks of the table but the xxx entry and find the ??? entry, which is the *Euler's method approximation for* $y(b)$.

Example 1.6.4. Using three steps of Euler's method, estimate $x(1)$, where $x''-3x'+2x = 1$, $x(0) = 0$, $x'(0) = 1$

First, we rewrite $x'' - 3x' + 2x = 1$, $x(0) = 0$, $x'(0) = 1$, as a system of first-order differential equations with initial conditions. Let $x_1 = x$, $x_2 = x'$, so

$$\begin{aligned} x_1' &= x_2, & x_1(0) &= 0, \\ x_2' &= 1 - 2x_1 + 3x_2, & x_2(0) &= 1. \end{aligned}$$

This is the differential equation rewritten as a system in standard form. (In general, the tabular method applies to any system but it must be in standard form.)

Taking $h = (1-0)/3 = 1/3$, we have

t	x_1	$x_2/3$	x_2	$(1-2x_1+3x_2)/3$
0	0	1/3	1	4/3
1/3	1/3	7/9	7/3	22/9
2/3	10/9	43/27	43/9	xxx
1	73/27	xxx	xxx	xxx

So $x(1) = x_1(1) \sim 73/27 = 2.7\ldots$.

Here is one way to do this using Sage.

Sage

```
sage: RR = RealField(sci_not=0, prec=4, rnd='RNDU')
sage: t, x, y = PolynomialRing(RR,3,"txy").gens()
sage: f = y; g = 1-2*x+3*y
sage: L = eulers_method_2x2(f,g,0,0,1,1/3,1,method="none")
sage: L
[[0, 0, 1], [1/3, 0.35, 2.5], [2/3, 1.3, 5.5],
 [1, 3.3, 12], [4/3, 8.0, 24]]
sage: eulers_method_2x2(f,g, 0, 0, 1, 1/3, 1)
 t          x           h*f(t,x,y)        y          h*g(t,x,y)
 0          0             0.35            1            1.4
 1/3        0.35          0.88            2.5          2.8
 2/3        1.3           2.0             5.5          6.5
 1          3.3           4.5             12           11
sage: P1 = list_plot([[p[0],p[1]] for p in L])
sage: P2 = line([[p[0],p[1]] for p in L])
sage: show(P1+P2)
```

The plot of the approximation to $x(t)$ is given in Figure 1.16.

Next, we give a numerical version of Example 3.3.3.

Example 1.6.5. Consider

$$\begin{cases} x' &= -4y, \quad x(0) = 400, \\ y' &= -x, \quad y(0) = 100. \end{cases}$$

Use two steps of Euler's method to approximate $x(1/2)$ and $y(1/2)$.

t	x	$hf(t,x,y) = -2y$	y	$hg(t,x,y) = -x/2$
0	400	-100	100	-100
1/4	300	0	0	-75
1/2	300	75	-75	-75

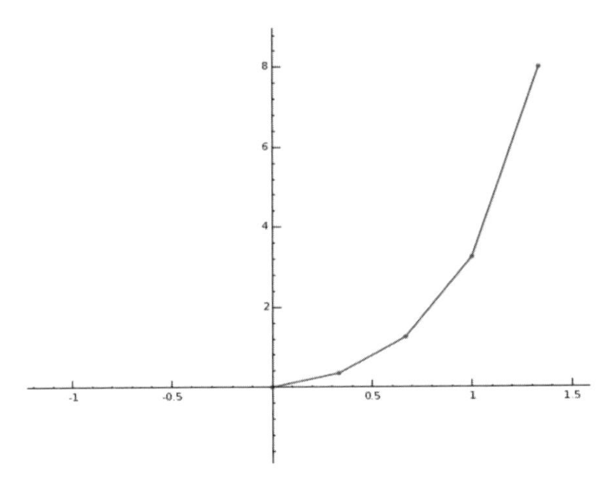

Figure 1.16: Euler's method with $h = 1/3$ for $x'' - 3x' + 2x = 1$, $x(0) = 0$, $x'(0) = 1$.

This computation of the approximations $x(1/2) \approx 300$, $y(1/2) \approx -75$ suggests that the y values lose some time between $t = 0.25$ and $t = 0.5$. This suggestion is further supported by the following more refined approximation:

Sage

```
sage: eulers_method_2x2(f,g, 0, 400, 100, 1/10, 1/2)
      t                x                  h*f(t,x,y)           y             h*g(t,x,y)
      0               400                 -40.00              100          -40.00
   1/10               360.0               -24.00              60.00             -36.00
    1/5               336.0                -9.600             24.00             -33.60
   3/10               326.4                 3.840             -9.600            -32.64
    2/5               330.2                16.90             -42.24             -33.03
    1/2               347.1                30.11             -75.27             -34.71
```

This computation yields the approximations $x(1/2) \approx 347.1$, $y(1/2) \approx -42.24$. The Example 3.3.3 will tell us the exact answer.

Exercises

1. (a) Use Euler's method to estimate $x(1)$ if $x(0) = 1$ and $\frac{dx}{dt} = x + t^2$, using one, two, and four steps.

   (b) Find the exact value of $x(1)$ by solving the ODE (hint: it is a linear ODE).

   (c) Use the improved Euler's method to estimate $x(1)$ if $x(0) = 1$ and $\frac{dx}{dt} = x + t^2$, using one, two, and four steps.

2. Use Sage and Euler's method with $h = 1/3$ to find the approximate values of $x(1)$ and $y(1)$, where

$$\begin{cases} x' = x + y + t, & x(0) = 0, \\ y' = x - y, & y(0) = 0, \end{cases}$$

3. Use two steps of the improved Euler's method to find the approximate value of $x(1)$ where $x' = x^2 + t^2$, $x(0) = 1$.

4. Use Euler's method with $h = 1/10$ to find the approximate value of $x(1)$ where $x'' + x = 1$, $x(0) = 1$, $x'(0) = -1$.

1.7 Numerical solutions II: Runge-Kutta and other methods

The methods of §1.6 are sufficient for computing the solutions of many problems, but often we are given difficult cases that require more sophisticated methods. One class of methods are called the Runge-Kutta methods, particularly the fourth-order method of that class since it achieves a popular balance of efficiency and simplicity. Another class, the multistep methods, use information from some number m of previous steps. Within that class, we will briefly describe the Adams-Bashforth method. Finally, we will say a little bit about adaptive step sizes—i.e., changing h adaptively depending on some local estimate of the error.

1.7.1 Fourth-order Runge-Kutta method

To explain why we describe the method as "fourth order" it is necessary to introduce a convenient (*big-O*) notation for the size of the errors in our approximations. We say that

$$f(x) = O(g(x))$$

as $x \to 0$ if there exist positive constants M and x_0 such that for all $x \in [-x_0, x_0]$ we have $|f(x)| \leq M|g(x)|$. The constant M is called the *implied constant.* This notation is also commonly used for the case when $x \to \infty$, but in this text we will always be considering the behavior of $f(x)$ near $x = 0$. For example, $\sin(x) = O(x)$ as $x \to 0$, but $\sin(x) = O(1)$ as $x \to \infty$. If $g(x)$ is asymptotic to 0 as $x \to 0$, then, roughly speaking, a function $f(x)$ is $O(g(x))$ if f approaches 0 at a rate equal to or faster than $g(x)$. As another example, the function $f(x) = 3x^2 + 6x^4$ is $O(x^2)$ (as $x \to 0$). As we approach $x = 0$, the higher order terms become less and less important, and eventually the $3x^2$ term dominates the value of the function. There are many concrete choices of x_0 and the implied constant M. One choice for this example would be $x_0 = 1$ and $M = 9$.

For a numerical method we are interested in how fast the error decreases as we reduce the step size. By "the error" we mean the global truncation error: for the problem of approximating $y(c)$ if y satisfies (1.9), the *global truncation error* is defined as $E(h) = |y(c) - y_n|$. Here y_n is our approximate value for y at $x = c$ after taking n steps of step size $h = \frac{c-a}{n}$.

The global truncation error ignores the rounding error that is inevitable using floating-point arithmetic. We ignore rounding error for two reasons: it is usually not a big problem, and it is implementation specific so it is hard to make general statements about it. Rounding error increases with the number of steps taken and can be important if there are near-cancellations between terms of the slope function.

For Euler's method, $E(h) = O(h)$, and we say that it is a first-order method. This means that as we decrease the step size h, at some point our error will become linear in h. In other words, we expect that if we halve the step size our error will be reduced by half. The improved Euler method is a second-order method, so $E(h) = O(h^2)$. This is very good news, because while the improved Euler method involves roughly twice as much work per step as the Euler method, the error will eventually fall quadratically in h.

The *fourth-order Runge-Kutta method* involves computing four slopes and taking a weighted average of them. We denote these slopes as k_1, k_2, k_3, and k_4, and the formulas are

$$\begin{aligned} x_{n+1} &= x_n + h, \\ y_{n+1} &= y_n + h(k_1 + 2k_2 + 2k_3 + k_4)/6, \end{aligned}$$

where

$$k_1 = f(x_n, y_n),$$

$$k_2 = f(x_n + h/2, y_n + hk_1/2),$$

$$k_3 = f(x_n + h/2), y_n + hk_2/2),$$

and

$$k_4 = f(x_n + h, y_n + hk_3).$$

Example 1.7.1. Let us consider the IVP $y' = \frac{y(y-x)}{x(y+x)}$, $y(1) = 1$; suppose we wish to approximate $y(2)$. Table 1.3 compares the Euler, improved Euler, and fourth-order Runge-Kutta (RK4) approximations for various numbers of steps from 1 to 512.

steps	Euler	Improved Euler	RK4
1	1.0	0.916666666667	0.878680484793
2	0.933333333333	0.889141488073	0.876938215214
4	0.90307164531	0.880183944727	0.876770226006
8	0.889320511452	0.877654079757	0.876757721415
16	0.882877913323	0.87698599324	0.87675688939
32	0.879775715551	0.876814710289	0.876756836198
64	0.878255683243	0.876771374145	0.876756832844
128	0.877503588678	0.876760476927	0.876756832634
256	0.877129540678	0.876757744807	0.876756832621
512	0.876943018826	0.876757060805	0.876756832620

Table 1.3: Some comparisons

The final Runge-Kutta value is correct to the number of digits shown. Note that even after 512 steps, Euler's method has not acheived the accuracy of four steps of Runge-Kutta or 16 steps of the improved Euler method.

1.7.2 Multistep methods: Adams-Bashforth

The idea of multistep methods is to use a sequence of solution or slope values to extrapolate to the next value. One of the most popular is the Adams-Bashforth method. To start such methods the first few values must be found by some other means, such as a Runge-Kutta method.

The *fourth-order Adams-Bashforth method* is

$$\begin{aligned} x_{n+1} &= x_n + h, \\ y_{n+1} &= y_n + \tfrac{h}{24}(55f_n - 59f_{n-1} + 37f_{n-2} - 9f_{n-3}), \end{aligned} \tag{1.11}$$

where $f_i = f(x_i, y_i)$.

Exercise

1.* Write a Sage function that implements the Adams-Bashforth method. To "prime" the method it is necessary to generate four initial solution values, for which you should use the fourth-order Runge-Kutta method. Compare the accuracy of your function at $x = \pi$ to the fourth-order Runge-Kutta method for the IVP

$$y' = y + \sin(x) \qquad y(0) = -1/2,$$

by first computing the exact value of $y(\pi)$.

1.7.3 Adaptive step size

In our discussion of numerical methods we have only considered a fixed step size. In some applications this is sufficient, but usually it is better to adaptively change the step size to keep the local error below some tolerance. One approach for doing this is to use two different methods for each step, and if the methods differ by more than the tolerance we decrease the step size. The Sage code below implements this for the improved Euler method and the fourth-order Runge-Kutta method.

Sage

```
def RK24(xstart, ystart, xfinish, f, nsteps = 10,  tol = 10^(-5.0)):
    '''
    Simple adaptive step-size routine. This compares the improved
    Euler method and the fourth-order Runge-Kutta method to
    estimate the error.

    EXAMPLE:
    The exact solution to this IVP is y(x) = exp(x), so y(1)
    should equal e = 2.718281828...
```

```
    Initially the stepsize is 1/10 but this is decreased during
    the calculation:
        sage: esol = RK24(0.0,1.0,1.0,lambda x,y: y)
        sage: print "Error is: ", N(esol[-1][1]-e)
        Error is:  -8.66619043193850e-9
    '''
    sol = [ystart]
    xvals = [xstart]
    h = N((xfinish-xstart)/nsteps)
    while xvals[-1] < xfinish:
        # Calculate slopes at various points:
        k1 = f(xvals[-1],sol[-1])
        rk2 = f(xvals[-1] + h/2,sol[-1] + k1*h/2)
        rk3 = f(xvals[-1] + h/2,sol[-1] + rk2*h/2)
        rk4 = f(xvals[-1] + h,sol[-1] + rk3*h)
        iek2 = f(xvals[-1] + h,sol[-1] + k1*h)
        # Runge-Kutta increment:
        rk_inc = h*(k1+2*rk2+2*rk3+rk4)/6
        # Improved Euler increment:
        ie_inc = h*(k1+iek2)/2
        #Check if the error is smaller than the tolerance:
        if abs(ie_inc - rk_inc) < tol:
            sol.append(sol[-1] + rk_inc)
            xvals.append(xvals[-1] + h)
        # If not, halve the stepsize and try again:
        else:
            h = h/2
    return zip(xvals,sol)
```

More sophisticated implementations will also increase the step size when the error stays small for several steps. A very popular scheme of this type is the Runge-Kutta-Fehlberg method, which combines fourth- and fifth-order Runge-Kutta methods in a particularly efficient way [A-ode].

Exercises.

1. For the initial value problem $y(1) = 1$ and $\frac{dy}{dx} = \frac{y}{x}\left(\frac{y-x}{x+y}\right)$, approximate $y(2)$ by using a

   (a) two step improved Euler method

   (b) one step fourth-order Runge-Kutta method.

2. Use Euler's method and the fourth-order Runge-Kutta method to estimate $x(1)$ if $x(0) = 1$ and $\frac{dx}{dt} = x + t^2$, using two steps. For this question you can use a calculator but you should write out the steps explicitly.

3. Use all three of the numerical methods (Euler, improved Euler, and fourth-order Runge-Kutta) in Sage to find the value of $y(1)$ to four digits of accuracy using the smallest possible number of steps if $y' = xy - y^3$ and $y(0) = 1$.

4. Let
$$\operatorname{sinc}(x) = \frac{\sin(x)}{x}.$$
Compare the results of using the fourth-order Runge-Kutta and Adams-Bashforth methods to approximate $y(3)$ for the IVP $\frac{dy}{dx} = \operatorname{sinc}(x) - y$, $y(0) = 1$ for 5, 10, and 100 steps. Use the Runge-Kutta values to prime the Adams-Bashforth method.

5* Modify the adaptive step size code so that the step size is increased if the estimated error is below some threshold. Pick an initial value problem and time your code compared to the original version.

6* Sometimes we wish to model quantities that are subjected to random influences, as in Brownian motion or in financial markets. If the random influences change on arbitrarily small time scales, the result is usually nondifferentiable. In these cases, it is better to express the evolution in terms of integrals, but they are often called *stochastic differential equations* nonetheless. One example from financial markets is a model of the value of a stock, which in integral form is
$$S(t) = S_0 + \int_0^t \mu S \, ds + \int_0^t \sigma S \, dW,$$
where W is a Brownian motion. For such a model there is a stochastic analog of Euler's method called the *Euler-Maruyama method* (see [H-sde] for more details and references). The main subtlety is correctly scaling the increment of the Brownian motion with the step size: $dW = \sqrt{dt}\, w_{0,1}$, where $w_{0,1}$ is a sample from a normal distribution with mean 0 and standard deviation 1. So for this example the Euler-Maruyama method gives
$$S_{i+1} = S_i + \mu S_i h + \sigma \sqrt{h} S_i w_{0,1}.$$
Implement the Euler-Maruyama method to simulate an IVP of this example with $\mu = 2$ and $\sigma = 1$ from $t = 0$ to $t = 1$ with 100 steps, with $S(0) = S_0 = 1$. To generate the $w_{0,1}$ in `Sage` you can use the `normalvariate` command. Compute the average trajectory of 100 simulation runs—is it equal to the deterministic ODE $S' = \mu S$? (This is a reasonable guess since the expected value of the Brownian motion W is 0.)

1.8 Newtonian mechanics

We briefly recall how the physics of the falling body problem leads naturally to a differential equation (this was already mentioned in §1.1 and forms a part of Newtonian mechanics [M-mech]). Consider a mass m falling due to gravity. We orient coordinates so that downward is positive. Let $x(t)$ denote the distance the mass has fallen at time t and $v(t)$ its velocity at time t. We assume that only two forces act: the force due to gravity, F_{grav}, and the force due to air resistance, F_{res}. In other words, we assume that the total force is given by

$$F_{total} = F_{grav} + F_{res}.$$

We know that $F_{grav} = mg$, where $g > 0$ is the gravitational constant, from high school physics. We assume, as is common in physics, that air resistance is proportional to velocity: $F_{res} = -kv = -kx'(t)$, where $k \geq 0$ is a constant. Newton's second law [N-mech] tells us that $F_{total} = ma = mx''(t)$. Putting these all together gives $mx''(t) = mg - kx'(t)$, or

$$v'(t) + \frac{k}{m}v(t) = g. \tag{1.12}$$

This is the differential equation governing the motion of a falling body. Equation (1.12) can be solved by various methods: separation of variables or by integrating factors. If we assume $v(0) = v_0$ is given and if we assume $k > 0$ then the solution is

$$v(t) = \frac{mg}{k} + \left(v_0 - \frac{mg}{k}\right)e^{-kt/m}. \tag{1.13}$$

In particular, we see that the limiting velocity is $v_{limit} = \frac{mg}{k}$.

Example 1.8.1. Wile E. Coyote (see [W-mech] if you haven't seen him before) has mass 100 kg (with chute). The chute is released 30 seconds after the jump from a height of 2000 m. The force due to air resistance is given by $\vec{F}_{res} = -k\vec{v}$, where

$$k = \begin{cases} 15, & \text{chute closed,} \\ 100, & \text{chute open.} \end{cases}$$

Find

(a) the distance and velocity functions during the time when the chute is closed (i.e., $0 \leq t \leq 30$ s);

(b) the distance and velocity functions during the time when the chute is open (i.e., $30 \leq t$ s);

(c) the time of landing;

(d) the velocity of landing. (Does Wile E. Coyote survive the impact?)

Solution. Taking $m = 100$, $g = 9.8$, $k = 15$, and $v(0) = 0$ in (1.13), we find

$$v_1(t) = \frac{196}{3} - \frac{196}{3}e^{-\frac{3}{20}t}.$$

This is the velocity with the time t starting the moment the parachutist jumps. After $t = 30$ seconds, this reaches the velocity $v_0 = \frac{196}{3} - \frac{196}{3}e^{-9/2} = 64.607\ldots$. The distance fallen is

$$\begin{aligned} x_1(t) &= \int_0^t v_1(u)\,du \\ &= \frac{196}{3}t + \frac{3920}{9}e^{-\frac{3}{20}t} - \frac{3920}{9}. \end{aligned}$$

After 30 s, the body has fallen $x_1(30) = \frac{13720}{9} + \frac{3920}{9}e^{-9/2} = 1529.283\ldots$ m.

Taking $m = 100$, $g = 9.8$, $k = 100$, and $v(0) = v_0$, we find

$$v_2(t) = \frac{49}{5} + e^{-t}\left(\frac{833}{15} - \frac{196}{3}e^{-9/2}\right).$$

This is the velocity with the time t starting the moment with which Wile E. Coyote opens his chute (i.e., 30 seconds after jumping). The distance fallen is

$$\begin{aligned} x_2(t) &= \int_0^t v_2(u)\,du + x_1(30) \\ &= \tfrac{49}{5}\,t - \tfrac{833}{15}\,e^{-t} + \tfrac{196}{3}\,e^{-t}e^{-9/2} + \tfrac{71099}{45} + \tfrac{3332}{9}\,e^{-9/2}. \end{aligned}$$

Now let us solve this using Sage.

Sage

```
sage: RR = RealField(sci_not=0, prec=50, rnd='RNDU')
sage: t = var('t')
sage: v = function('v', t)
sage: m = 100; g = 98/10; k = 15
sage: de = lambda v: m*diff(v,t) + k*v - m*g
sage: desolve(de(v),[v,t],[0,0])
196/3*(e^(3/20*t) - 1)*e^(-3/20*t)
sage: soln1 = lambda t: 196/3-196*exp(-3*t/20)/3
sage: P1 = plot(soln1(t),0,30,plot_points=1000)
sage: RR(soln1(30))
64.607545559502
```

This solves for the velocity before the coyote's chute is opened, $0 < t < 30$. The last number is the velocity Wile E. Coyote is traveling at the moment he opens his chute.

Sage

```
sage: t = var('t')
sage: v = function('v', t)
sage: m = 100; g = 98/10; k = 100
sage: de = lambda v: m*diff(v,t) + k*v - m*g
sage: desolve(de(v),[v,t],[0,RR(soln1(30))])
1/10470*(102606*e^t + 573835)*e^(-t)
sage: soln2 = lambda t: 1/10470*(102606*e^t + 573835)*e^(-t)
sage: RR(soln2(0))
64.607545367718
sage: RR(soln1(30))
64.607545559502
sage: P2 = plot(soln2(t),30,50,plot_points=1000)
sage: show(P1+P2, dpi=300)
```

This solves for the velocity after the coyote's chute is opened, $t > 30$. The last command plots the velocity functions together as a single graph.[7] The terms at the end of `soln2` were added to ensure $x_2(30) = x_1(30)$.

Next, we find the distance traveled at time t.

Sage

```
sage: x1 = integral(soln1(t),t) - 3920/9
sage: x1(t=0)
0
sage: RR(x1(30))
1529.2830296034
```

This solves for the distance the coyote traveled before the chute was open, $0 < t < 30$. The last number says that he has gone about 1965 m when he opens his chute.

Sage

```
sage: x2 = integral(soln2(t),t)+114767/2094
sage: x2(t=0)
0
sage: RR(x2(t=42.44)+x1(t=30))
2000.0025749711
sage: P4 = plot(x2(t),30,50)
sage: show(P3+P4)
```

(Again, you see a break in the graph because of the round-off error.) The terms at the end of `x2` were added to ensure $x_2(30) = x_1(30)$. You know he is close to the ground at $t = 30$ and going quite fast (about 65 m/s!). It makes sense that he will hit the ground soon afterward (with a large puff of smoke, if you've seen the cartoons), even though his chute will have slowed him down somewhat.

The graph of the velocity $0 < t < 50$ is in Figure 1.17. Notice how it drops at $t = 30$ when the chute is opened.

The time of impact is $t_{impact} = 42.4\ldots$. This was found numerically by solving $x_2(t) = 2000 - x_1(30)$ (using Sage).

The velocity of impact is $v_2(t_{impact}) \approx 9.8$ m/s.

Exercises

[7] You would see a break in the graph if you omitted the Sage's `plot` option `plot_points=1000`. That is because the number of samples taken of the function by default is not sufficient to capture the jump the function takes at $t = 30$.

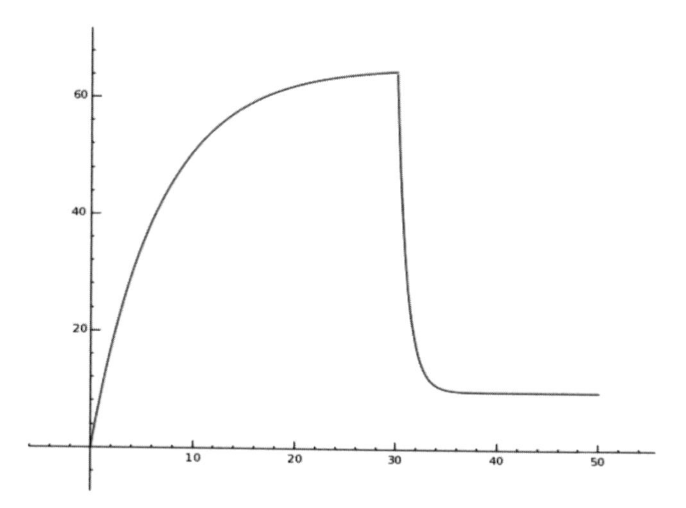

Figure 1.17: Velocity of falling parachutist.

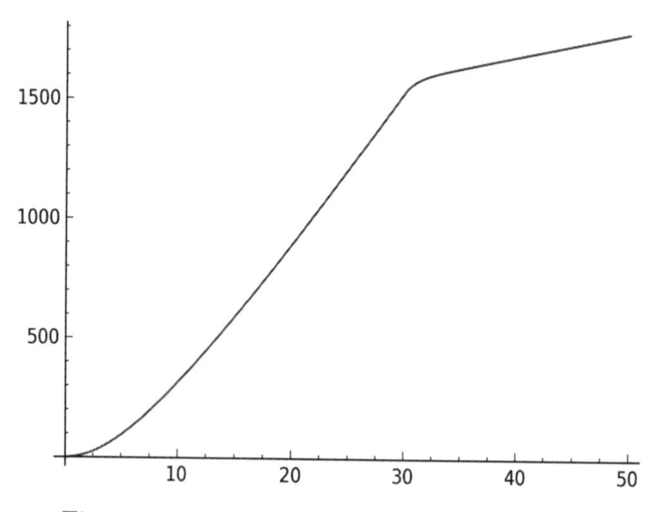

Figure 1.18: Distance fallen by a parachutist.

1. Drop an object with mass 10 kg from a height of 2000 m. Suppose the force due to air resistance is given by $\vec{F}_{res} = -10\vec{v}$. Find the velocity after 10 s using `Sage`. Plot this velocity function for $0 < t < 10$.

2. Drop an object with mass 100 kg from a height of 1000 m. Suppose the force due to air resistance is given by $\vec{F}_{res} = -5\vec{v}$. Find the time of impact. Find the velocity of impact.

1.9 Application to mixing problems

Suppose that we have two chemical substances where one is soluble in the other, such as salt and water. Suppose that a mixture of these substances is poured into a tank and the resulting well-mixed solution pours out through a valve at the bottom. (The term "well-mixed" is used to indicate that the fluid being poured in is assumed to instantly dissolve into a homogeneous mixture the moment it goes into the tank.) The rough idea is depicted in Figure 1.19.

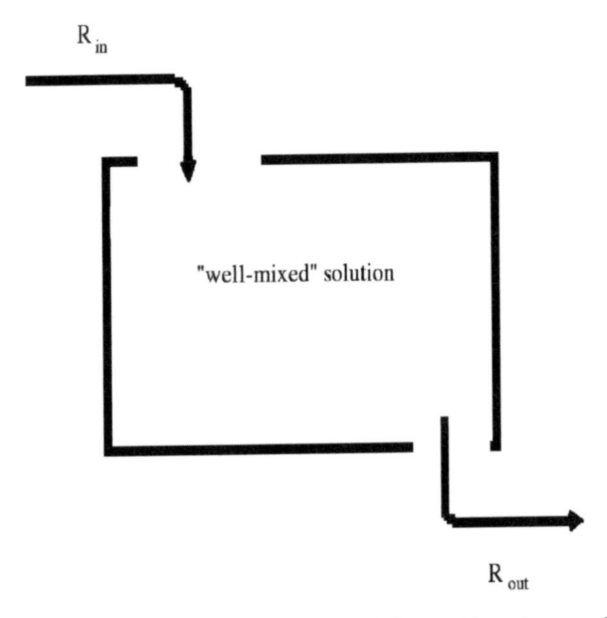

Figure 1.19: Solution pours into a tank, mixes with another type of solution, and then pours out.

Assume for concreteness that the chemical substances are salt and water. Let

- $A(t)$ denote the amount of salt at time t,
- FlowRateIn = the rate at which the solution pours into the tank,
- FlowRateOut = the rate at which the mixture pours out of the tank,
- C_{in} = "concentration in" = the concentration of salt in the solution being poured into the tank,
- C_{out} = "concentration out" = the concentration of salt in the solution being poured out of the tank,
- R_{in} = rate at which the salt is being poured into the tank = (FlowRateIn)(C_{in}),

- R_{out} = rate at which the salt is being poured out of the tank = (FlowRateOut)(C_{out}).

Remark 1.9.1. Some things to make note of:

- If FlowRateIn = FlowRateOut then the "water level" of the tank stays the same.
- We can determine C_{out} as a function of other quantities:

$$C_{out} = \frac{A(t)}{T(t)},$$

where $T(t)$ denotes the volume of solution in the tank at time t.

- The rate of change of the amount of salt in the tank, $A'(t)$, more properly could be called the "net rate of change." If you think of it this way then you see that $A'(t) = R_{in} - R_{out}$.

Now the differential equation for the amount of salt arises from the above equations:

$$A'(t) = (\text{FlowRateIn})C_{in} - (\text{FlowRateOut})\frac{A(t)}{T(t)}.$$

Example 1.9.1. Consider a tank with 200 l of salt-water solution, 30 g of which is salt. Pouring into the tank is a brine solution at a rate of 4 l/min and with a concentration of 1 g/l. The well-mixed solution pours out at a rate of 5 l/min. Find the amount of salt at time t.

We know that

$$A'(t) = (\text{FlowRateIn})C_{in} - (\text{FlowRateOut})\frac{A(t)}{T(t)} = 4 - 5\frac{A(t)}{200-t}, \quad A(0) = 30.$$

Writing this in the standard form $A' + pA = q$, we have

$$A(t) = \frac{\int \mu(t)q(t)\,dt + C}{\mu(t)},$$

where $\mu = e^{\int p(t)\,dt} = e^{-5\int \frac{1}{200-t}\,dt} = (200-t)^{-5}$ is the "integrating factor." This gives $A(t) = 200 - t + C \cdot (200-t)^5$, where the initial condition implies $C = -170 \cdot 200^{-5}$.

Here is one way to do this using Sage:

Sage

```
sage: t = var('t')
sage: A = function('A', t)
sage: de = lambda A: diff(A,t) + (5/(200-t))*A - 4
sage: desolve(de(A),[A,t])
(t - 200)^5*(c - 1/(t - 200)^4)
```

This is the form of the general solution. Let us now solve this general solution for c, using the initial conditions.

Sage

```
sage: c,t = var('c,t')
sage: tank = lambda t: 200-t
sage: solnA = lambda t: (c + 1/tank(t)^4)*tank(t)^5
sage: solnA(t)
(c - (1/(t - 200)^4))*(t - 200)^5
sage: solnA(0)
-320000000000*(c - 1/1600000000)
sage: solve([solnA(0) == 30],c)
[c == 17/32000000000]
sage: c = 17/32000000000
sage: solnA(t)
(17/32000000000 - (1/(t - 200)^4))*(t - 200)^5
sage: P = plot(solnA(t),0,200)
sage: show(P)
```

This plot is given in Figure 1.20.

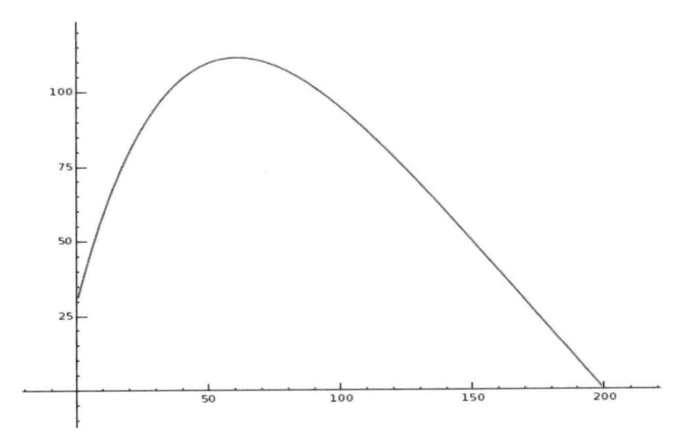

Figure 1.20: $A(t)$, $0 < t < 200$, $A' = 4 - 5A(t)/(200 - t)$, $A(0) = 30$.

Exercises.

1. Now use Sage to solve the same problem but with the same flow rate out as 4 l/min (note: in this case, the "water level" in the tank is constant). Find and plot the solution $A(t)$, $0 < t < 200$.

2. Consider a tank containing 1000 l of brine with 100 kg of salt dissolved. Pure water is added to the tank at a rate of 10 l/s, and stirred mixture is drained out at a rate of 10 L/s. Find the time at which only 1 kg of salt is left in the tank.

3. Consider two water tanks that are linked in a cascade, i.e. the first tank empties into the second. Suppose the first tank has 100 l of water in it, and the second has 300 l of water. Each tank initially has 50 kg of salt dissolved in the water. Suppose that pure water flows into the first tank at 5 l per minute, well-mixed water flows from the first tank to the second at the same rate (5 l/min), and well-mixed water also flows out of the second tank at 5 l/min.

   (a) Find the amount of salt in the first tank, $x_1(t)$. Note that this does not depend on what is happening in the second tank.

   (b) Find the amount of salt in the second tank, $x_2(t)$.

   (c) Find the time when there is the maximum amount of salt in the second tank.

4. A 120 gal tank initially contains 90 lb of salt dissolved in 90 gal of water. Brine containing 2 lb/gal of salt flows into the tank at the rate of 4 gal/min and the well-stirred mixture flows out of the tank at the rate of 3 gal/min. A differential equation for the amount x (in pounds) of salt in the tank is

   (a) $\frac{dx}{dt} + \frac{1}{30}x = 8$; (b) $\frac{dx}{dt} + \frac{3}{90+2t}x = 4$; (c) $\frac{dx}{dt} + \frac{3}{90+t}x = 8$; (d) $\frac{dx}{dt} + 90x = 2$.

5. A tank contains 100 gal of water with 100 lb of salt dissolved in it. Saltwater with a concentration of 4 lb/gal is pumped into the tank at a rate of 1 gal/min. The well-stirred mixture is pumped out at the rate of 2 gal/min. At what time does the tank contain the largest amount of salt? How much salt is in the tank at that time?

6. A tank contains 100 gal of water with 100 lb of salt dissolved in it. Saltwater with a concentration of 4 lb/gal is pumped into the tank at a rate of 1 gal/min. The well-stirred mixture is pumped out at the rate of 2 gal/min.

   (a) Find the amount, in pounds, of salt in the tank as a function of time.

   (b) At what time does the tank contain the largest amount of salt?

7. Curly, Moe, and Larry are incompetent butlers for Daffy Duck, who orders them to prepare his bath. The bathtub holds 100 gal when filled. Curly begins pouring in a soap-water solution at a rate of 1/2 gal/min. The soap concentration is 1/4 lb/gal in his soap-water solution. Larry, in charge of putting the plug in the drain, screwed up and now the bath water is pouring down the drain at a rate of 1/2 gal/min. Originally, Moe filled the bath full of pure water.

   (a) What is the differential equation modeling the mixing problem for the amount $A(t)$ of soap at time t?

   (b) Find the amount (in pounds) of soap in the bathtub after 200 min.

1.10 Application to cooling problems

Place an object, such as a cup of coffee, in a room. How does its temperature behave? Convection is described by *Newton's law of cooling*, which states that the rate of change of the temperature of an object is proportional to its relative temperature (i.e., the difference in temperature between the object and its surroundings).

Figure 1.21: How fast does a cup of coffee cool?

If

- $T = T(t)$ is the temperature of an object at time t,
- T_{room} is the ambient temperature of the object's surroundings,

then

$$T'(t) = k \cdot (T - T_{room}), \tag{1.14}$$

where k is a constant of proportionality. If T_{room} is a constant then the solution has the form

$$T(t) = T_{room} + Ae^{kt}, \tag{1.15}$$

where A is a constant. This can be derived by the method of separation of variables. It can also be verified directly by plugging (1.15) into (1.14).

Example 1.10.1. Sherlock Holmes is awoken by a phone call from a policeman at 3:30 a.m. A body has been discovered and foul play is suspected. Sherlock tells the police to determine the temperature of the body and, when he arrives at the scene 45 minutes later, he takes the temperature again. The two readings that cold 60° F morning were 80° F and 70° F.

Figure 1.22: Holmes has a new case!

First, write down the differential equation governing the temperature:

$$T'(t) = k \cdot (T - 60),$$

since $T_{room} = 60$.

Next, we find the time of death.[8]

Here is one way to do this using Sage:

Sage

```
sage: t,k,Troom = var("t,k,Troom")
sage: T = function("T", t)
sage: de = diff(T, t) == k*(T - Troom)
sage: desolve(de, dvar=T, ivar=t)
(Troom*e^(-k*t) + c)*e^(k*t)
sage: Troom = 60
sage: Tk = desolve(de, dvar=T, ivar=t); Tk
(Troom*e^(-k*t) + c)*e^(k*t)
sage: c = var("c")
sage: solve(Tk(t=0)==80, c)
[c == -Troom + 80]
sage: Troom = 60; c = -Troom + 80
sage: solve(Tk(t=45)==70, k)
[k == log((-Troom/c + 70/c)^(1/45)*e^(2/45*I*pi)),
..., k == 1/45*log(-Troom/c + 70/c)]
sage: k = 1/45*log(-Troom/c + 70/c); RR(k)
-0.0154032706791099
sage: k
1/45*log(1/2)
sage: Tt = Tk(Troom = 60, c = -Troom + 80, k = 1/45*log(1/2))
sage: Tt
20*(3*e^(-1/45*t*log(1/2)) + 1)*e^(1/45*t*log(1/2))
```

[8]We use 98.6° F for the normal body temperature.

Therefore,

$$T(t) = 20(3(1/2)^{-t/45} + 1)(1/2)^{t/45} = 60 + 20(1/2)^{t/45}.$$

Exercises.

1. Verify (1.15).

2. The following is a somewhat more general heating and cooling DE that includes additional heating and air conditioning of a building.

$$\frac{dT}{dt} = K[M(t) - T(t)] + H(t) + U(t). \tag{1.16}$$

   (a) Suppose that $M(t) = 75^\circ + 20\sin(\pi t/12)$, $K = 2$, $H(t) = 75^\circ$, $U(t) = 0$. By use of `Sage`, find and plot solutions, where the initial condition is

      (a) $T(0) = 60^\circ$,

      (b) $T(0) = 70^\circ$,

      (c) $T(0) = 80^\circ$ over a 24-hour period.

   (b) What do you observe about the effect of the initial condition on the solution as time progresses?

   (c) Solve (1.16) explicitly and determine what factor in the solution might account for this behavior?

   (d) Let $T(0) = 60^\circ$ and let the other values stay fixed except let $K = 2,\ 4,\ 6$. Graph the three solutions for these three K values.

   Can you describe the effect of the time constant K on the solutions?

3. A cup of coffee at 120° F is placed in a room of constant temperature 70° F. After 10 minutes it cools to 80° F.

   (a) What is the ODE modeling the temperature of the coffee at time t?

   (b) How long did it take to cool to 100° F? (Leave your answer in symbolic form.)

4. Sherlock Holmes is awoken by a phone call from a policeman at 4 a.m. A body has been discovered and foul play is suspected. Sherlock tells the police to determine the temperature of the body and, when he arrives at the scene 30 minutes later, he takes the temperature again. The two readings that cold 50° F morning were 90° F and 85° F. Find the time of death (use 98.6° F for the normal body temperature).

5. A glass of cola at 50° F is placed in a room of constant temperature 70° F. After 10 min it warms to 60° F.

(a) What is the differential equation modeling the temperature of the coffee at time t?

(b) How long did it take to cool to 65° F?

Chapter 2

Second-order differential equations

If people do not believe that mathematics is simple, it is only because they do not realize how complicated life is.

—*John von Neumann*

2.1 Linear differential equations

2.1.1 Solving homogeneous constant-coefficient ODEs

The problem we want to address in this section is to solve a differential equation of the form

$$a_n x^{(n)} + \cdots + a_1 x' + a_0 x = 0, \tag{2.1}$$

where the a_i's are given constants and $x = x(t)$ is our unknown dependent variable. Such a differential equation expression is a "homogeneous" (since there is no term depending only on the independent variable t), "linear" (since we are not operating on the dependent variable y except in "legal" ways), "constant-coefficient" (since the coefficients of the derivatives of x are all constants, i.e., not depending on t) differential equation.

Remark 2.1.1. Sometimes it is convenient to rewrite the above differential equation using *symbolic notation*:

replace	by
x'	Dx
x''	D^2x
x'''	D^3x
$x^{(n)}$	D^nx

For example, the nonhomogeneous constant-coefficient linear second-order differential equation

$$x'' + 2x' + 7x = \sin(t)$$

would be written in symbolic notation as

$$(D^2 + 2D + 7)x = D^2x + 2Dx + 7x = \sin(t).$$

In general, we write (2.1) in symbolic notation as follows:

$$a_nD^nx + \cdots + a_1Dx + a_0x = 0.$$

To solve the homogeneous differential equation in (2.1), please *memorize these rules:*

real $r \Rightarrow e^{rt}$
complex $\alpha \pm \beta i \Rightarrow e^{\alpha t}\cos(\beta t),\ e^{\alpha t}\sin(\beta t)$
repeated roots $\Rightarrow$ repeated solutions, boosted each time by t

The solution process

- Find the characteristic polynomial of the equation.
- Find the roots of the polynomial.
- Construct the solution by interpreting the root according to the rules above.

Example 2.1.1. Find the general solution to $y''' - 2y'' - 3y' = 0$.

Solution. The characteristic polynomial is $r^3 - 2r^2 - 3r$. Factor this into $r(r+1)(r-3)$, so the roots are 0, -1, 3. The solution is

$$y = c_1e^{0t} + c_2e^{-1t} + c_3e^3t = c_1 + c_2e^{-t} + c_3e^{3t}.$$

The following fact is called the fundamental theorem of algebra. It is not easy to prove. The first proof was given by Carl F. Gauss in the 1800s.

Theorem 2.1.1. A polynomial of degree n has n roots—if you count complex roots and repeated roots.

Each of these special cases gets special handling by these rules. This also means you should always end up with n independent solutions to an nth-degree equation.

Example 2.1.2. Find the general solution to $y'' - 4y' + 40y = 0$.

The characteristic polynomial is $r^2 - 4r + 40$. The roots are (using the quadratic formula:)

$$r = \frac{-(-4) \pm \sqrt{4^2 - 4(1)(40)}}{2(1)} = \frac{4 \pm \sqrt{-144}}{2} = 2 \pm 6i.$$

The solution is

$$y = c_1e^{2t}\cos(6t) + c_2e^{2t}\sin(6t).$$

For a repeated root, repeat the solution as many times, but "boost" the solution (multiply by t) each time it is repeated.

Example 2.1.3. Find the general solution to $y''' + 6y'' + 12y' + 8y = 0$.

The characteristic polynomial is $r^3+6r^2+12r+8$. This factors into $(r+2)^3$, so the roots are $r = -2, -2, -2$ (repeated three times). The solution (with boosting) is

$$y = c_1 e^{-2t} + c_2 te^{-2t} + c_3 t^2 e^{-2t}.$$

Example 2.1.4. Find the general solution to $y^{(4)} + 8y'' + 16y = 0$.

The characteristic polynomial is r^4+8r^2+16. This factors into $(r^2+4)^2 = (r+2i)^2(r-2i)^2$. This is a conjugate pair of complex roots repeated twice. The solution (with boosting) is

$$\begin{aligned} y &= c_1 e^{0t}\cos(2t) + c_2 e^{0t}\sin(2t) + c_3 te^{0t}\cos(2t) + c_4 te^{0t}\sin(2t) \\ &= c_1\cos(2t) + c_2\sin(2t) + c_3 t\cos(2t) + c_4 t\sin(2t). \end{aligned}$$

Some more quick examples are listed in Table 2.1 to allow for displaying more cases. The left-hand column represents the roots of the characteristic polynomial of the homogeneous, linear, constant-coefficient differential equation. The right-hand column represents the general solution to that differential equation, where the c_i's are constants.

Roots:	General solution
$1, 2, 3$	$c_1e^t + c_2e^{2t} + c_3e^{3t}$
$1, 2, 2$	$c_1e^t + c_2e^{2t} + c_3te^{2t}$
$2, 2, 2$	$c_1e^{2t} + c_2te^{2t} + c_3t^2e^{2t}$
$3, 3, 0, 0, 0, 0$	$c_1e^{3t} + c_2te^{3t} + c_3 + c_4t + c_5t^2 + c_6t^3$
$1 \pm 7i, 3 \pm 2i$	$c_1e^t\cos(7t) + c_2e^t\sin(7t) + c_3e^{3t}\cos(2t) + c_4e^{3t}\sin(2t)$
$1 \pm 7i, 1 \pm 7i$	$c_1e^t\cos(7t) + c_2e^t\sin(7t) + c_3te^t\cos(7t) + c_4te^t\sin(7t)$
$0, 1, -1, \pm i$	$c_1 + c_2e^t + c_3e^{-t} + c_4\cos t + c_5\sin t$

Table 2.1: Some examples

Example 2.1.5. Consider the differential equation

$$y''' - 3y'' + 3y' + y = 0.$$

At the present time, `Sage` uses `Maxima` to solve all such differential equations. We use `Sympy` (included with `Sage`) to solve this[1]:

[1] `Sympy` is an independent computer algebra system, written in Python. Go to its website `http://sympy.org/` for more information.

Sympy
```
sage: x = Symbol("x")
sage: y = Function('y')
sage: dsolve(Derivative(y(x),x,x,x)-3*Derivative(y(x),x,x)+3*Derivative(y(x),x)-y(x), y(x))
y(x) == (C1 + C2*x + C3*x**2)*exp(x)
```

The solution is, according to Sympy,

$$y = c_1 e^x + c_2 x e^x + c_3 x^2 e^x.$$

This agrees with the solution obtained by the process described above.

Another example of this method is given in §2.3.

Exercises

1. Use Euler's formula ($e^{ia} = \cos(a) + i\sin(a)$) to calculate (a) $(e^{\alpha+\beta i} + e^{\alpha-\beta i})/2$, (b) $(e^{\alpha+\beta i} - e^{\alpha-\beta i})/(2i)$. Simplify your answers as much as possible.

2. Write the general solution corresponding to the given set of roots.

   (a) $0, 0, 2, 2 \pm 4i$ (b) $6 \pm i, 6 \pm i, 10$
   (c) $4, -7, -7, 1 \pm 5i$ (d) $1 \pm 8i, \pm 5i$
   (e) $0, 1, 1, 1, 1, -9$ (f) $2, 3 \pm 5i, 3 \pm 5i, 3 \pm 5i$

3. Write the set of roots corresponding to the given general solution.

   (a) $c_1 e^{4t} + c_2 t e^{4t} + c_3 e^{-8t}$ (b) $c_1 \cos(5t) + c_2 \sin(5t) + c_3 e^{10t}$
   (c) $c_1 + c_2 t + c_3 t^2 + c_4 e^{-t} + c_5 t e^{-t}$ (d) $c_1 e^{4t}\cos t + c_2 e^{4t}\sin t + c_3 t e^{4t}\cos t + c_4 t e^{4t}\sin t$
   (e) $c_1 e^{2t} + c_2 e^{-5t} + c_3 t e^{-5t}$ (f) $c_1 e^{4t} + c_2 + c_3\cos(10t) + c_4\sin(10t)$

4. Find the general solution to the given differential equation.

   (a) $y''' - 2y'' + y' = 0$

   (b) $y^{(4)} + 9y'' = 0$

   (c) $y'' - 9y' + 20y = 0$.

5. Solve the initial value problem.

   (a) $y'' - 5y' + 4y = 0$, $y(0) = 0$, $y'(0) = -3$

   (b) $y'' - 2y' + 5y = 0$, $y(0) = 1$, $y'(0) = -2$.

6. Solve the initial value problem $y'' - 4y = 0$, $y(0) = 4$, $y'(0) = 2$ given that $y_1 = e^{2x}$ and $y_2 = e^{-2x}$ are both solutions to the ODE.

7. (a) Find the general solution to $x'' - x' - 20x = 0$.

(b) Use the general solution to solve the initial value problem $x'' - x' - 20x = 0$, $x(0) = 1$, $x'(0) = 1$.

8. Find the general solution to $y'' + 6y' = 0$.

9. Find the general solution to $4y'' + 4y' + y = 0$.

10. Find the general solution to $y'' + 10y' + 25y = 0$.

11. Solve the initial value problem $y'' - 6y' + 25y = 0$, $y(0) = 6$, $y'(0) = 2$.

12. For what second-order constant-coefficient linear homogeneous ODE would $y = C_1 + C_2x$ be the general solution?

13. Find the general solution to $y^{(4)} - 6y^{(3)} + 9y'' = 0$.

14. Find the general solution of $6y^{(4)} + 5y^{(3)} + 18y'' + 20y' - 24y = 0$ given that $y = \cos(2x)$ is a solution. (Hint: use polynomial division on the characteristic polynomial.)

2.2 Linear differential equations, revisited

To begin, we want to describe the general form a solution to a linear ordinary differential equation can take. We want to describe the solution as a sum of terms which can be computed explicitly in some way.

Before doing this, we introduce two pieces of terminology.

- Suppose $f_1(t)$, $f_2(t)$, ..., $f_n(t)$ are given functions. A *linear combination* of these functions is another function of the form

$$c_1f_1(t) + c_2f_2(t) + \cdots + c_nf_n(t),$$

for some constants c_1, ..., c_n. For example, $3\cos(t) - 2\sin(t)$ is a linear combination of $\cos(t)$, $\sin(t)$. An "arbitrary linear combination" of $\cos(t)$, $\sin(t)$ would be written as $c_1 \cos(t) + c_2 \sin(t)$, where c_1, c_2 are constants (which could be also considerd as parameters).

- A linear differential equation of the form

$$y^{(n)} + b_1(t)y^{(n-1)} + \cdots + b_{n-1}(t)y' + b_n(t)y = f(t), \tag{2.2}$$

is called *homogeneous* if $f(t) = 0$ (i.e., f is the 0 function) and otherwise it is called *nonhomogeneous*.

Consider the nth order differential equation

$$y^{(n)} + b_1(t)y^{(n-1)} + \cdots + b_{n-1}(t)y' + b_n(t)y = 0. \tag{2.3}$$

Suppose there are n functions $y_1(t)$, ..., $y_n(t)$ such that

- each $y = y_i(t)$ $(1 \leq i \leq n)$ satisfies this homogeneous differential equation (2.3),
- *every* solution y to (2.3) is a linear combination of these functions $y_1, \ldots, y_n$:

$$y = c_1 y_1 + \cdots + c_n y_n,$$

for some (unique) constants $c_1, \ldots, c_n$.

In this case, the y_i's are called *fundamental solutions.* This generalizes the discussion of fundamental solutions for second-order differential equations in §1.3.2 above.

Remark 2.2.1. If you are worrying that this definition is not very practical, then don't. There is a simple condition (the "Wronskian test") which will make it much easier to see if a collection of n functions forms a set of fundamental solutions. See Theorem 1.3.4 or Theorem 2.3.1.

The following result describes the general solution to a linear differential equation.

Theorem 2.2.1. Consider a linear differential equation of the above form (2.2), for some given continuous functions $b_1(t)$, ..., $b_n(t)$ and $f(t)$. Then the following hold.

- There are n functions $y_1(t)$, ..., $y_n(t)$ (the above-mentioned fundamental solutions), each satisfying the homogeneous differential equation, such that every solution y_h to (2.3) can be written

$$y_h = c_1 y_1 + \cdots + c_n y_n, \tag{2.4}$$

for some constants $c_1, \ldots, c_n$. (This function y_h is sometimes called the *homogeneous part* or the *complementary part* of the solution to (2.2).)

- Suppose you know a solution $y_p(t)$ to (2.2). Then every solution $y = y(t)$ to the differential equation (2.2) has the form

$$y(t) = y_h(t) + y_p(t), \tag{2.5}$$

where y_h is as above. (The function y_p is sometimes called the *particular solution* to (2.2) and y is called the *general solution.*)

- Conversely, every function of the form (2.5), for any constants c_i as in (2.4), for $1 \leq i \leq n$, is a solution to (2.2).

Example 2.2.1. Recall Example 1.1.6 where we looked for functions solving $x' + x = 1$ by "guessing." We found that the function $x_p(t) = 1$ is a particular solution to $x' + x = 1$. The function $x_1(t) = e^{-t}$ is a fundamental solution to $x' + x = 0$. The general solution is therefore $x(t) = 1 + c_1 e^{-t}$, for a constant c_1.

Example 2.2.2. Let's look for functions solving $x'' - x = 1$ by guessing. Motivated by the above example, we find that the function $x_p(t) = -1$ is a particular solution to $x'' - x = 1$. The functions $x_1(t) = e^t$, $x_2(t) = e^{-t}$ are fundamental solutions to $x'' - x = 0$. The general solution is therefore $x(t) = 1 + c_1 e^{-t} + c_2 e^t$, for constants c_1 and c_2.

Example 2.2.3. The charge on the capacitor of an RLC electrical circuit[2] is modeled by a second-order linear differential equation (see, for example, [C-linear] for details).

Series RLC circuit notations:

- $E = E(t)$—the voltage of the power source (a battery or other "electromotive force," measured in volts, V);
- $q = q(t)$—the current in the circuit (measured in coulombs, C);
- $i = i(t)$—the current in the circuit (measured in amperes, A);
- L—the inductance of the inductor (measured in henrys, H);
- R—the resistance of the resistor (measured in ohms, Ω);
- C—the capacitance of the capacitor (measured in farads, F).

The charge q on the capacitor satisfies the linear IVP:

$$Lq'' + Rq' + \frac{1}{C}q = E(t), \quad q(0) = q_0, \quad q'(0) = i_0. \tag{2.6}$$

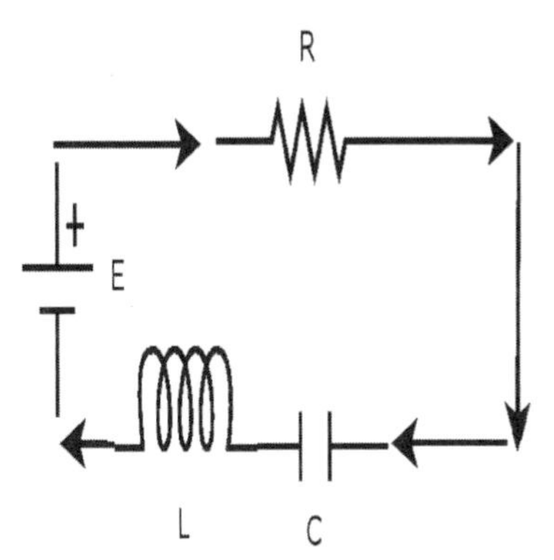

Figure 2.1: RLC circuit.

This circuit example is analogous to the spring example (Example 2.2.4).

[2]The abbreviation "RLC" stands for a simple closed electrical circuit, as in Figure 2.1, which has a resistor, an inductor, and a capacitor.

Example 2.2.4. The displacement from equilibrium of a mass attached to a spring suspended from a ceiling as in Figure 2.2 is modeled by a 2-nd order linear differential equation [O-ivp].

Spring-mass notations:

- $f(t)$—the external force acting on the spring (if any);
- $x = x(t)$—the displacement from equilibrium of a mass attached to a spring;
- m—the mass;
- b—the damping constant (if, say, the spring is immersed in a fluid);
- k—the spring constant.

The displacement x satisfies the linear IVP:

$$mx'' + bx' + kx = f(t), \quad x(0) = x_0, \quad x'(0) = v_0. \tag{2.7}$$

Figure 2.2: Spring-mass model.

Notice that each general solution to an nth order differential equation has n "degrees of freedom" (the arbitrary constants c_i). According to Theorem 2.2.1, to find the general solution of a linear differential equation, we need only find a particular solution y_p and n fundamental solutions $y_1(t), \ldots, y_n(t)$ ($n = 2$) to the differential equation in (2.7).

Example 2.2.5. Let us try to solve

$$x' + x = 1, \quad x(0) = c,$$

where $c = 1$, $c = 2$, and $c = 3$. (Three different IVPs, three different solutions; find each one.)

The first problem, $x' + x = 1$ and $x(0) = 1$, is easy. The solution to the differential equation $x' + x = 1$ which we guessed at in the previous example, $x(t) = 1$, satisfies this.

The second problem, $x' + x = 1$ and $x(0) = 2$, is not so simple. To solve this, we would like to know what the form of the general solution is. This would help use solve the IVP, as we did in §1.2. According to Theorem 2.2.1, the general solution x has the form $x = x_p + x_h$. In this case, $x_p(t) = 1$ and $x_h(t) = c_1 x_1(t) = c_1 e^{-t}$, by the earlier Example 1.1.6. Therefore, every solution to the differential equation above is of the form $x(t) = 1 + c_1 e^{-t}$, for some constant c_1. We use the initial condition to solve for c_1:

- $x(0) = 1$: $1 = x(0) = 1 + c_1 e^0 = 1 + c_1$ so $c_1 = 0$ and $x(t) = 1$ (as we saw above).
- $x(0) = 2$: $2 = x(0) = 1 + c_1 e^0 = 1 + c_1$ so $c_1 = 1$ and $x(t) = 1 + e^{-t}$.
- $x(0) = 3$: $3 = x(0) = 1 + c_1 e^0 = 1 + c_1$ so $c_1 = 2$ and $x(t) = 1 + 2e^{-t}$.

Here is one way[3] to use Sage to solve for c_1. We use Sage to solve the last IVP discussed above and then to plot the solution.

Sage

```
sage: t = var('t')
sage: x = function('x',t)
sage: desolve(diff(x,t)+x==1,[x,t])
(c + e^t)*e^(-t)
sage: c = var('c')
sage: solnx = lambda t: 1+c*exp(-t)  # the soln from desolve
sage: solnx(0)
c + 1
sage: solve([solnx(0) == 3],c)
[c == 2]
sage: c0 = solve([solnx(0) == 3], c)[0].rhs()
sage: solnx1 = lambda t: 1+c0*exp(-t); solnx1(t)
sage: P = plot(solnx1(t),0,5)
sage: show(P)
```

This plot is shown in Figure 2.3.

Exercises

1. Let $q_1(t) = e^{-t}\cos(t)$ and $q_2(t) = e^{-t}\sin(t)$. Check that these are fundamental solutions to $q'' + 2q' + 2q = 0$. In particular, using the Wronskian, check that they are linearly independent.

[3] Of course, you can do this yourself, but this shows you the Sage syntax for solving equations. Type `solve?` in Sage to get more details.

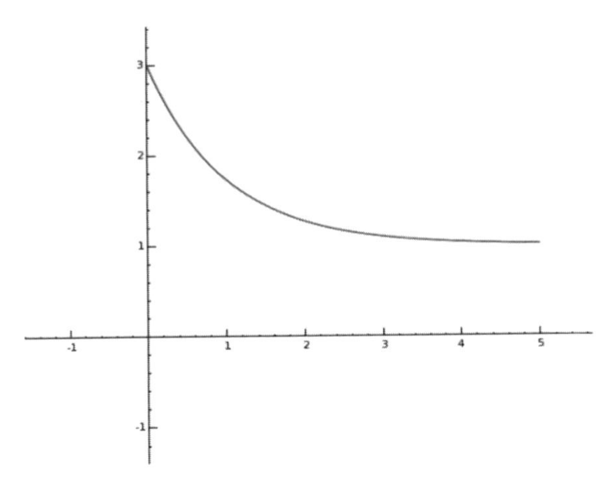

Figure 2.3: Solution to IVP $x' + x = 1$, $x(0) = 3$.

2. Let $x_1(t) = e^{-t}$ and $x_2(t) = te^{-t}$. Check that these are fundamental solutions to $x'' + 2x' + x = 0$. In particular, using the Wronskian, check that they are linearly independent.

3. (a) A 32 lb weight is attached to a spring having spring constant $k = 49$ with no damping. The mass is initially released with a 3 ft/s upward velocity 2 ft below equilibrium. (Please choose your orientation so that down is positive.) Find the differential equation and initial conditions for the displacement $x = x(t)$ of this spring-mass system.

   (b) A 32 lb weight is attached to a spring having spring constant $k = 101$ with damping constant $b = 20$ and an external force of $f(t) = 10$. The mass is initially released from rest at equilibrium. (Please choose your orientation so that down is positive.) Find the IVP for the displacement $x = x(t)$ of this spring-mass system.

4. A simple circuit with inductor $L = 1$ H, $R = 6$ Ω resistor, capacitor $C = 1/10$ F, and an $e(t) = 1$ V emf has initial charge and current both equal to 0. Find the differential equation and initial conditions modeling this circuit. Draw a circuit diagram representing this situation.

5. A simple circuit with inductor $L = 1$ H, no resistor, no electro-motive force, and capacitor $C = 1/100$ F, has initial charge -1 C and current equal to 1 A. Find the IVP modeling this circuit.

2.3 Linear differential equations, continued

To better describe the form a solution to a linear ODE can take, we need to better understand the nature of fundamental solutions and particular solutions.

Recall that the general solution to

$$y^{(n)} + b_1(t)y^{(n-1)} + \cdots + b_{n-1}(t)y' + b_n(t)y = f(t)$$

has the form $y = y_p + y_h$, where y_h is a linear combination of fundamental solutions to the corresponding homogeneous differential equation. Suppose we are also given n initial conditions $y(x_0) = a_0$, $y'(x_0) = a_1$, $\ldots$, $y^{(n-1)}(x_0) = a_{n-1}$. We want to be able to solve this nth-order initial value problem and obtain a unique solution. Indeed, these n conditions will nail down the arbitrary constants given in the linear combination defining y_h, so that y is unique.

Example 2.3.1. The general solution to the differential equation

$$x'' - 5x' + 6x = 0$$

has the form $x = x(t) = c_1 \exp(2t) + c_2 \exp(3t)$, for arbitrary constants c_1 and c_2. For example, suppose we impose the initial conditions $x(0) = 3$ and $x'(0) = 4$. (Of course, no matter what $x(0)$ and $x'(0)$ are given, we want to be able to solve for the coefficients c_1, c_2 in $x(t) = c_1 \exp(2t) + c_2 \exp(3t)$ to obtain a unique solution.) In this case, we obtain

$$3 = x(0) = c_1 + c_2, \quad 4 = x'(0) = 2c_1 + 3c_2.$$

Solving these two equations in two unknowns is easy: multiply the first equation by 2 and subtract from the second equation to obtain $-2 = c_2$. This forces $c_1 = 5$, so

$$x = 5e^{2t} - 2e^{3t}.$$

We shall solve this in a slightly different way below.

Based on the example above, a few questions arise.

Remark 2.3.1.

- How do we know this can be done?
- How do we know that (for some c_1, c_2) some linear combination $x(t) = c_1 \exp(2t) + c_2 \exp(3t)$ isn't identically 0 (which, if true, would imply that $x = x(t)$ couldn't possibly satisfy $x(0) = 3$ and $x'(0) = 4$)?

The complete answer to these questions actually involves methods from linear algebra which go beyond the scope of this book. The basic idea is not hard to understand, and it involves what is called the Wronskian [W-linear].

Before we motivate the idea of the Wronskian by returning to the above example, we need to recall a basic fact from linear algebra.

Lemma 2.3.1. *(Cramer's rule)* Consider the system of two equations in two unknowns x, y:

$$ax + by = s_1, \qquad cx + dy = s_2.$$

The solution to this system is

$$x = \frac{\det \begin{pmatrix} s_1 & b \\ s_2 & d \end{pmatrix}}{\det \begin{pmatrix} a & b \\ c & d \end{pmatrix}}, \qquad y = \frac{\det \begin{pmatrix} a & s_1 \\ c & s_2 \end{pmatrix}}{\det \begin{pmatrix} a & b \\ c & d \end{pmatrix}}.$$

Note the determinant $\det \begin{pmatrix} a & b \\ c & d \end{pmatrix} = ad - bc$ is in the denominator of both expressions. In particular, if the determinant is 0 then the formula is invalid (and in that case, the solution either does not exist or is not unique).

Example 2.3.2. Write the general solution to $x'' - 5x' + 6x = 0$ as $x(t) = c_1 x_1(t) + c_2 x_2(t)$ (we know that $x_1(t) = \exp(2t)$, $x_2(t) = \exp(3t)$, but we leave it in this more abstract notation to make a point). Assume the initial conditions $x(0) = 3$ and $x'(0) = 4$ hold. We can try solving for c_1, c_2 but plugging $t = 0$ into the general solution gives

$$3 = x(0) = c_1 e^0 + c_2 e^0 = c_1 + c_2, \qquad 4 = x'(0) = c_1 2e^0 + c_2 3e^0 = 2c_1 + 3c_2.$$

You can solve these by hand for c_1, c_2 (and you are encouraged to do so). However, to motivate Wronskians we shall use the initial conditions in the more abstract form of the general solution:

$$3 = x(0) = c_1 x_1(0) + c_2 x_2(0), \qquad 4 = x'(0) = c_1 x_1'(0) + c_2 x_2'(0).$$

Cramers' rule gives us the solution for this system of two equations in two unknowns c_1, c_2:

$$c_1 = \frac{\det \begin{pmatrix} 3 & x_2(0) \\ 4 & x_2'(0) \end{pmatrix}}{\det \begin{pmatrix} x_1(0) & x_2(0) \\ x_1'(0) & x_2'(0) \end{pmatrix}}, \qquad y = \frac{\det \begin{pmatrix} x_1(0) & 3 \\ x_1'(0) & 4 \end{pmatrix}}{\det \begin{pmatrix} x_1(0) & x_2(0) \\ x_1'(0) & x_2'(0) \end{pmatrix}}.$$

What you see in the denominator of these expressions is the Wronskian of the fundamental solutions x_1, x_2 evaluated at $t = 0$.

From the example above we see that Wronskians arise naturally in the process of solving for c_1 and c_2.

In general terms, what is a Wronskian? It is best to explain what this means not just for two functions, but for any finite number of functions. This more general case would be useful in case we wanted to try to solve a higher-order ODE by the same method. If $f_1(t)$, $f_2(t)$, $\ldots$, $f_n(t)$ are given n-times-differentiable functions then their *fundamental matrix* is the $n \times n$ matrix

$$\Phi = \Phi(f_1, \ldots, f_n) = \begin{pmatrix} f_1(t) & f_2(t) & \ldots & f_n(t) \\ f_1'(t) & f_2'(t) & \ldots & f_n'(t) \\ \vdots & \vdots & & \vdots \\ f_1^{(n-1)}(t) & f_2^{(n-1)}(t) & \ldots & f_n^{(n-1)}(t) \end{pmatrix}.$$

The determinant of the fundamental matrix is called the *Wronskian*, denoted $W(f_1, \ldots, f_n)$. The Wronskian actually helps us answer both questions in Remark 2.3.1 simultaneously.

Example 2.3.3. Take $f_1(t) = \sin^2(t)$, $f_2(t) = \cos^2(t)$, and $f_3(t) = 1$. Sage allows us to easily compute the Wronskian:

Sage
```
sage: t = var('t')
sage: Phi = matrix([[sin(t)^2,cos(t)^2,1],
            [diff(sin(t)^2,t),diff(cos(t)^2,t),0],
            [diff(sin(t)^2,t,t),diff(cos(t)^2,t,t),0]])
sage: Phi
[    sin(t)^2                    cos(t)^2                    1]
[    2*sin(t)*cos(t)            -2*sin(t)*cos(t)             0]
[   -2*sin(t)^2 + 2*cos(t)^2     2*sin(t)^2 - 2*cos(t)^2     0]
sage: det(Phi)
0
```

Here `det(Phi)` is the determinant of the fundamental matrix `Phi`. Since it is zero, this means

$$W(\sin(t)^2, \cos(t)^2, 1) = 0.$$

Let's try another example using Sage.

Sage
```
sage: t = var('t')
sage: Phi = matrix([[sin(t)^2,cos(t)^2], [diff(sin(t)^2,t),diff(cos(t)^2,t)]])
sage: Phi

[          sin(t)^2                cos(t)^2         ]
[ 2*cos(t)*sin(t)         -2*cos(t)*sin(t)        ]
sage: Phi.det()
-2*cos(t)*sin(t)^3 - 2*cos(t)^3*sin(t)
```

This means $W(\sin(t)^2, \cos(t)^2) = -2\cos(t)\sin(t)^3 - 2\cos(t)^3\sin(t)$, which is nonzero.

If there are constants $c_1, \ldots, c_n$, not all zero, for which

$$c_1 f_1(t) + c_2 f_2(t) \cdots + c_n f_n(t) = 0, \quad \text{for all } t, \tag{2.8}$$

then the functions f_i $(1 \leq i \leq n)$ are called *linearly dependent*. If the functions f_i $(1 \leq i \leq n)$ are not linearly dependent then they are called *linearly independent* (this definition is frequently seen for linearly independent vectors [L-linear] but holds for functions as well). This condition (2.8) can be interpreted geometrically as follows. Just as $c_1 x + c_2 y = 0$ is a line through the origin in the plane and $c_1x + c_2y + c_3z = 0$ is a plane containing the origin in 3-space, the equation

$$c_1x_1 + c_2x_2 + \cdots + c_nx_n = 0,$$

is a "hyperplane" containing the origin in n-space with coordinates $(x_1, \ldots, x_n)$. This condition (2.8) says geometrically that the graph of the space curve $\vec{r}(t) = (f_1(t), \ldots, f_n(t))$ lies entirely in this hyperplane. If you pick n functions "at random" then they are "probably" linearly independent, because "random" space curves don't lie in a hyperplane. But certainly not all collections of functions are linearly independent.

Example 2.3.4. Consider just the two functions $f_1(t) = \sin^2(t)$, $f_2(t) = \cos^2(t)$. We know from the Sage computation in the example above that these functions are linearly independent.

Sage

```
sage: P = parametric_plot((sin(t)^2,cos(t)^2),0,5)
sage: show(P)
```

The Sage plot of this space curve $\vec{r}(t) = (\sin(t)^2, \cos(t)^2)$ is given in Figure 2.4. It is obviously not contained in a line through the origin, therefore making it geometrically clear that these functions are linearly independent.

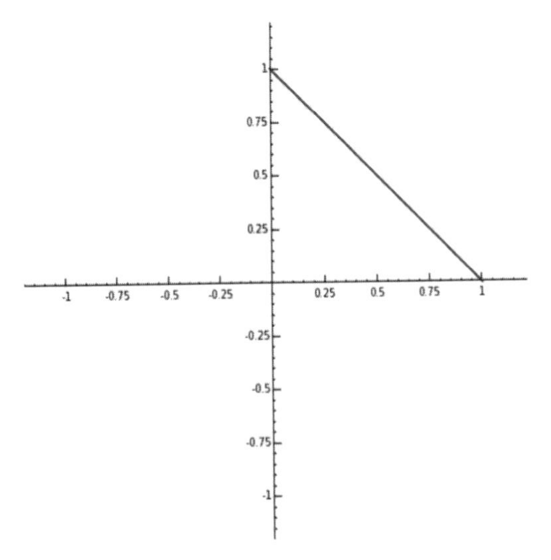

Figure 2.4: Parametric plot of $(\sin(t)^2, \cos(t)^2)$.

The following two results answer the questions in Remark 2.3.1.

Theorem 2.3.1. *(Wronskian test)* If $f_1(t)$, $f_2(t)$, ..., $f_n(t)$ are given n-times-differentiable functions with a non-zero Wronskian then they are linearly independent.

As a consequence of this theorem, and the Sage computation in the example above, $f_1(t) = \sin^2(t)$, $f_2(t) = \cos^2(t)$ are linearly independent.

Theorem 2.3.2. *Given any homogeneous nth linear ODE*

$$y^{(n)} + b_1(t)y^{(n-1)} + \cdots + b_{n-1}(t)y' + b_n(t)y = 0,$$

with differentiable coefficients, there always exists n solutions $y_1(t)$, ..., $y_n(t)$ which have a nonzero Wronskian.

The functions $y_1(t)$, ..., $y_n(t)$ in the above theorem are called *fundamental solutions*.

We shall not prove either of these theorems here. Please see [BD-intro] for further details.

Exercises

1. Use Sage to compute the Wronskian of

   (a) $f_1(t) = \sin(t)$, $f_2(t) = \cos(t)$;

   (b) $f_1(t) = 1$, $f_2(t) = t$, $f_3(t) = t^2$, $f_4(t) = t^3$;

   (c) $f_1(t) = \cos(2t)$, $f_2(t) = \sin(2t)$, $f_3(t) = 1 - \cos(2t)$.

2. Use the Wronskian test to check that

   (a) $y_1(t) = \sin(t)$, $y_2(t) = \cos(t)$ are fundamental solutions for $y'' + y = 0$;

   (b) $y_1(t) = 1$, $y_2(t) = t$, $y_3(t) = t^2$, $y_4(t) = t^3$ are fundamental solutions for $y^{(4)} = y'''' = 0$.

2.4 Undetermined coefficients method

The method of undetermined coefficients [U-uc] can be used to solve the following type of problem.

Problem

Solve

$$ay'' + by' + cy = f(x), \tag{2.9}$$

where $a \neq 0$, b and c are constants, and $f(x)$ is a special type of function. (Even the case $a = 0$ can be handled similarly, though some of the discussion below might need to be slightly modified.) The assumption that $f(x)$ is of a "special" form will be explained in more detail later.

More or less equivalent is the method of annihilating operators [A-uc] (they solve the same class of DEs), but that method will be discussed separately.

2.4.1 Simple case

For the moment, let us assume $f(x)$ has the "simple" form $a_1 \cdot p(x) \cdot e^{a_2 x} \cdot \cos(a_3 x)$, or $a_1 \cdot p(x) \cdot e^{a_2 x} \cdot \sin(a_3 x)$, where a_1, a_2, and a_3 are constants and $p(x)$ is a polynomial.

Solution

- Solve the homogeneous DE $ay'' + by' + cy = 0$ as follows. Let r_1 and r_2 denote the roots of the characteristic polynomial $aD^2 + bD + c = 0$.

  —$r_1 \neq r_2$ real: the solution is $y = c_1 e^{r_1 x} + c_2 e^{r_2 x}$.

  —$r_1 = r_2$ real: if $r = r_1 = r_2$ then the solution is $y = c_1 e^{rx} + c_2 x e^{rx}$.

  —r_1, r_2 complex: if $r_1 = \alpha + i\beta$, $r_2 = \alpha - i\beta$, where α and β are real, then the solution is $y = c_1 e^{\alpha x} \cos(\beta x) + c_2 e^{\alpha x} \sin(\beta x)$.

  Denote this solution y_h (some texts use y_c) and call this the *homogeneous part* of the solution. (Some texts call this the complementary part of the solution.)

- Compute $f(x)$, $f'(x)$, $f''(x)$, $\ldots$. Write down the list of all the different terms which arise (via the product rule), ignoring constant factors, plus signs, and minus signs:

$$t_1(x),\ t_2(x),\ \ldots,\ \ t_k(x).$$

  If any one of these agrees with y_1 or y_2 then multiply them all by x. (If, after this, any of them *still* agrees with y_1 or y_2 then multiply them all again by x.)

- Let y_p be a linear combination of these functions (your guess):

$$y_p = A_1 t_1(x) + \cdots + A_k t_k(x).$$

  This is called the *general form of the particular solution* (when you have *not* solved for the constants A_i). The A_i's are called *undetermined coefficients.*

- Plug y_p into (2.9) and solve for $A_1, \ldots, A_k$.

- Let $y = y_h + y_p = y_p + c_1 y_1 + c_2 y_2$. This is the *general solution* to (2.9). If there are any initial conditions for (2.9), solve for c_1, c_2 now.

Diagramatically:

Factor characteristic polynomial

$\downarrow$

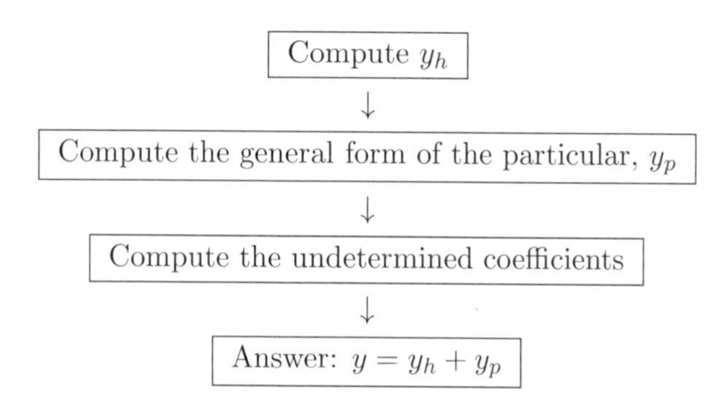

Example 2.4.1. Solve

$$y'' - y = \cos(2x).$$

- The characteristic polynomial is $r^2 - 1 = 0$, which has ± 1 for roots. The homogeneous solution is therefore $y_h = c_1 e^x + c_2 e^{-x}$.
- We compute $f(x) = \cos(2x)$, $f'(x) = -2\sin(2x)$, $f''(x) = -4\cos(2x)$, ... They are all linear combinations of

  $$f_1(x) = \cos(2x), \ \ f_2(x) = \sin(2x).$$

  None of these agrees with $y_1 = e^x$ or $y_2 = e^{-x}$, so we do not multiply by x.
- Let y_p be a linear combination of these functions:

  $$y_p = A_1 \cos(2x) + A_2 \sin(2x).$$

- You can compute both sides of $y_p'' - y_p = \cos(2x)$:

  $$(-4A_1 \cos(2x) - 4A_2 \sin(2x)) - (A_1 \cos(2x) + A_2 \sin(2x)) = \cos(2x).$$

  Equating the coefficients of $\cos(2x)$, $\sin(2x)$ on both sides gives two equations in two unknowns: $-5A_1 = 1$ and $-5A_2 = 0$. Solving, we get $A_1 = -1/5$ and $A_2 = 0$.
- The general solution: $y = y_h + y_p = c_1 e^x + c_2 e^{-x} - \frac{1}{5}\cos(2x)$.

Example 2.4.2. Solve

$$y'' - y = x\cos(2x).$$

- The characteristic polynomial is $r^2 - 1 = 0$, which has ± 1 for roots. The homogeneous solution is therefore $y_h = c_1 e^x + c_2 e^{-x}$.

- We compute $f(x) = x\cos(2x)$, $f'(x) = \cos(2x) - 2x\sin(2x)$, $f''(x) = -2\sin(2x) - 2\sin(2x) - 2x\cos(2x)$, ... They are all linear combinations of

$$f_1(x) = \cos(2x),\ f_2(x) = \sin(2x),\ f_3(x) = x\cos(2x),\ .f_4(x) = x\sin(2x).$$

None of these agrees with $y_1 = e^x$ or $y_2 = e^{-x}$, so we do not multiply by x.

- Let y_p be a linear combination of these functions:

$$y_p = A_1\cos(2x) + A_2\sin(2x) + A_3x\cos(2x) + A_4x\sin(2x).$$

- In principle, you can compute both sides of $y_p'' - y_p = x\cos(2x)$ and solve for the A_i's. (Equate coefficients of $x\cos(2x)$ on both sides, equate coefficients of $\cos(2x)$ on both sides, equate coefficients of $x\sin(2x)$ on both sides, and equate coefficients of $\sin(2x)$ on both sides. This gives four equations in four unknowns, which can be solved to obtain $A_1 = A_4 = 0$, $A_2 = 4/25$, and $A_3 = -1/5$.)

This can be confirmed using the following Sage commands.

Sage

```
sage: x = var("x")
sage: y = function("y", x)
sage: de = diff(y,x,2) - y == x*cos(x)
sage: desolve(de, y)
k1*e^x + k2*e^(-x) - 1/2*x*cos(x) + 1/2*sin(x)
```

Exercises

1. Find the general solution to $y'' + 6y' = e^t$.

2. Find the general solution to $4y'' + 4y' + y = 2$.

3. Find the general solution to $y'' + 10y' + 25y = 3\cos(t)$.

4. Solve the initial value problem $y'' - 6y' + 25y = 3t\sin(2t)$, $y(0) = 6$, $y'(0) = 2$.

2.4.2 Nonsimple case

More generally, suppose that you want to solve $ay'' + by' + cy = f(x)$, where $f(x)$ is a *sum* of functions of the "simple" functions in §2.4.1. In other words, $f(x) = f_1(x) + f_2(x) + \cdots + f_k(x)$, where each $f_j(x)$ is of the form $c \cdot p(x) \cdot e^{ax} \cdot \cos(bx)$, or $c \cdot p(x) \cdot e^{ax} \cdot \sin(bx)$, where a, b, c are constants and $p(x)$ is a polynomial. You can proceed in either one of the following ways.

1. Split up the problem by solving each of the k problems. Suppose

DE	has solution
$ay'' + by' + cy = f_1(x)$	$y = y_1(x)$
$ay'' + by' + cy = f_2(x)$	$y = y_2(x)$
...	...
$ay'' + by' + cy = f_k(x)$	$y = y_k(x)$

   The solution to $ay'' + by' + cy = f_1(x) + \cdots + f_k(x)$ is then $y = y_1 + y_2 + .. + y_k$ (the *superposition principle*).

2. Proceed as in the examples above but with the following slight revision:

   - Find the homogeneous solution y_h to $ay'' + by' = cy = 0$, $y_h = c_1 y_1 + c_2 y_2$.
   - Compute $f(x)$, $f'(x)$, $f''(x)$, ... Write down the list of all the different, individual terms that arise, ignoring constant factors, plus signs, and minus signs:

$$t_1(x),\ t_2(x),\ \ldots,\ \ t_k(x).$$

   - Group these terms into their *families*. Each family is determined from its parent(s) - which are the terms in $f(x) = f_1(x) + f_2(x) + \cdots + f_k(x)$ from which they arose by differentiation. For example, if $f(x) = x\cos(2x) + e^{-x}\sin(x) + \sin(2x)$ then the individual terms you get from differentiating this repeatedly (ignoring constants, plus signs, and minus signs) are

$$x\cos(2x), x\sin(2x), \cos(2x), \sin(2x) \qquad (\text{from } x\cos(2x)),$$

$$e^{-x}\sin(x), e^{-x}\cos(x) \qquad (\text{from } e^{-x}\sin(x)),$$

   and

$$\sin(2x), \cos(2x) \qquad (\text{from } \sin(2x)).$$

   The first group absorbs the last group, since you can count only the *different* terms. Therefore, there are only two families in this example:

$$\{x\cos(2x), x\sin(2x), \cos(2x), \sin(2x)\}$$

   is a "family" (with "parent" $x\cos(2x)$ and the other terms as its "children") and

$$\{e^{-x}\sin(x), e^{-x}\cos(x)\}$$

   is a "family" (with "parent" $e^{-x}\sin(x)$ and the other term as its "child").
   If any one of these terms agrees with y_1 or y_2 then multiply the *entire family* by x. In other words, if any child or parent is "bad" then the entire family is "bad." (If, after this, any of them *still* agrees with y_1 or y_2 then multiply them all again by x.)

- Let y_p be a linear combination of these functions (your guess):

$$y_p = A_1 t_1(x) + \cdots + A_k t_k(x).$$

This is called the *general form of the particular solution.* The A_i's are called *undetermined coefficients.*

- Plug y_p into (2.9) and solve for $A_1, \ldots, A_k$.
- Let $y = y_h + y_p = y_p + c_1 y_1 + c_2 y_2$. This is the *general solution* to (2.9). If there are any initial conditions for (2.9), solve for c_1, c_2 *last—after the undetermined coefficients.*

Example 2.4.3. Solve

$$y'' + 4y = x\cos(2x).$$

- The characteristic polynomial is $r^2+4=0$, which has $\pm 2i$ for roots. The homogeneous solution is therefore $y_h = c_1\cos(2x) + c_2\sin(2x)$.
- We compute $f(x) = x\cos(2x)$, $f'(x) = \cos(2x) - 2x\sin(2x)$, $f''(x) = -2\sin(2x) - 2\sin(2x) - 2x\cos(2x)$, ... They are all linear combinations of

$$f_1(x) = \cos(2x),\ f_2(x) = \sin(2x),\ f_3(x) = x\cos(2x),\ .f_4(x) = x\sin(2x).$$

Two of these agree with $y_1 = \cos(2x)$ or $y_2 = \sin(2x)$, so we multiply by x:

$$f_1(x) = x\cos(2x),\ f_2(x) = x\sin(2x),\ f_3(x) = x^2\cos(2x),\ .f_4(x) = x^2\sin(2x).$$

- Let y_p be a linear combination of these functions:

$$y_p = A_1 x\cos(2x) + A_2 x\sin(2x) + A_3 x^2\cos(2x) + A_4 x^2\sin(2x).$$

- You can compute both sides of $y_p'' + 4y_p = x\cos(2x)$ and solve for the A_i's. After a little algebra we obtain $A_1 = 1/16$, $A_2 = 0$, $A_3 = 0$, and $A_4 = 1/8$.
- The general solution is therefore

$$y = y_h + y_p = c_1\cos(2x) + c_2\sin(2x) + \frac{1}{16}x\cos(2x) + \frac{1}{4}x^2\sin(2x).$$

Example 2.4.4. Solve

$$y''' - y'' - y' + y = 12xe^x + 2.$$

We can use `Sage` for this.

Sage

```
sage: x = var("x")
sage: y = function("y",x)
sage: R.<D> = PolynomialRing(QQ[I], "D")
sage: p = D^3 - D^2 - D + 1
sage: p.factor()
 (D + 1) * (D - 1)^2
sage: p.roots()
 [(-1, 1), (1, 2)]
```

So the roots of the characteristic polynomial are $-1, 1, 1$, which means that the homogeneous part of the solution is

$$y_h = c_1e^x + c_2xe^x + c_3e^{-x}.$$

Sage

```
sage: de = lambda y: diff(y,x,3) - diff(y,x,2) - diff(y,x,1) + y
sage: c1, c2, c3 = var("c1, c2, c3")
sage: yh = c1*e^x + c2*x*e^x + c3*e^(-x)
sage: de(yh)
 0
```

This just confirmed that y_h solves $y''' - y'' - y' + y = 0$. Using Sage and the derivatives of $f(x) = 12xe^x + 2$, we generate the general form of the particular solution below.

Sage

```
sage: f = 12*x*e^x + 2
sage: diff(f,x,1); diff(f,x,2); diff(f,x,3)
 12*x*e^x + 12*e^x
 12*x*e^x + 24*e^x
 12*x*e^x + 36*e^x
sage: A1, A2, A3 = var("A1,A2,A3")
sage: yp = A1*x^2*e^x + A2*x^3*e^x + A3
```

Note that we multiplied part of this y_p by x so there is no "agreement" with y_h now. Plug this into the differential equation and compare coefficients of like terms to solve for the undertermined coefficients:

Sage

```
sage: de(yp)
```

```
 12*A2*x*e^x + 4*A1*e^x + 6*A2*e^x + A3
sage: solve([12*A2 == 12, 6*A2+4*A1 == 0, A3==2],A1,A2,A3)
 [[A1 == (-3/2), A2 == 1, A3 == 2]]
```

Finally, let's check if this is correct:

Sage

```
sage: y = yh + (-3/2)*x^2*e^x + (1)*x^3*e^x + 2
sage: de(y)
 12*x*e^x + 2
```

This can be confirmed using Sympy (used in Sage):

Sage

```
sage: x = Symbol("x")
sage: y = Function('y')
sage: de = y(x).diff(x,3) - y(x).diff(x,2) - y(x).diff(x) + y(x) - (12*x*e^x + 2)
sage: dsolve(de, y(x))
y(x) ==  C3*exp(-x) + (C1 + C2*x + x**3 - 3*x**2/2)*exp(x) + 2
```

Exercises

1. Using Sage, solve
$$y''' - y'' + y' - y = 12xe^x,$$
2. Using Sage, solve
$$y''' + y'' + y' + y = -3e^x + 10xe^x.$$
3. Solve these using Sympy, as above.
4. Consider the differential equation
$$x'' + 9x = 10t^2e^t + 2t\cos(3t).$$
Find the general form of a particular solution $x_p(t)$. Do not solve for the undetermined coefficients.
5. Consider the ODE
$$x' - 2x = 10t^{10}e^{2t}.$$
Find the general form of a particular solution $x_p(t)$. Do not solve for the undetermined coefficients.

2.5 Annihilator method

We consider again the same type of differential equation as in §2.4, but take a slightly different approach here. Recall that we refer to the differentiation operator as "D," that is $D = \frac{d}{dx}$.

Problem
Solve

$$ay'' + by' + cy = f(x). \tag{2.10}$$

We *assume* that $f(x)$ is of the form $c \cdot p(x) \cdot e^{ax} \cdot \cos(bx)$, or $c \cdot p(x) \cdot e^{ax} \cdot \sin(bx)$, where a, b, c are constants and $p(x)$ is a polynomial.
Solution

- Write the ODE in symbolic form $(aD^2 + bD + c)y = f(x)$.
- Find the homogeneous solution y_h to $ay'' + by' + cy = 0$, $y_h = c_1y_1 + c_2y_2$.
- Find the differential operator L which annihilates $f(x)$: $Lf(x) = 0$. The following *annihilator table* may help.

Function	Annihilator
x^k	D^{k+1}
$x^k e^{ax}$	$(D-a)^{k+1}$
$x^k e^{\alpha x}\cos(\beta x)$	$(D^2 - 2\alpha D + \alpha^2 + \beta^2)^{k+1}$
$x^k e^{\alpha x}\sin(\beta x)$	$(D^2 - 2\alpha D + \alpha^2 + \beta^2)^{k+1}$

- Find the general solution to the homogeneous ODE, $L \cdot (aD^2 + bD + c)y = 0$.
- Let y_p be the function you get by taking the solution you just found and subtracting from it any terms in y_h.
- Solve for the undetermined coefficients in y_p as in the method of undetermined coefficients.

Example 2.5.1. Solve

$$y'' - y = \cos(2x).$$

- The DE is $(D^2 - 1)y = \cos(2x)$.
- The characteristic polynomial is $r^2 - 1 = 0$, which has ± 1 for roots. The homogeneous part of the solution is therefore $y_h = c_1e^x + c_2e^{-x}$.
- We find that $L = D^2 + 4$ annihilates $\cos(2x)$.

- We solve $(D^2+4)(D^2-1)y=0$. The roots of the characteristic polynomial $(r^2+4)(r^2-1)$ are $\pm 2i, \pm 1$. The solution is

$$y = A_1\cos(2x) + A_2\sin(2x) + A_3e^x + A_4e^{-x}.$$

- This solution agrees with y_h in the last two terms, so we choose

$$y_p = A_1\cos(2x) + A_2\sin(2x).$$

- Now solve for A_1 and A_2 as before: Compute both sides of $y_p'' - y_p = \cos(2x)$,

$$(-4A_1\cos(2x) - 4A_2\sin(2x)) - (A_1\cos(2x) + A_2\sin(2x)) = \cos(2x).$$

  Next, equate the coefficients of $\cos(2x)$ and $\sin(2x)$ on both sides to get two equations in two unknowns. Solving, we get $A_1 = -1/5$ and $A_2 = 0$.

- The general solution: $y = y_h + y_p = c_1e^x + c_2e^{-x} - \frac{1}{5}\cos(2x)$.

Exercises

1. Find the general solution to $x'' - x' = t$.

2. Consider the differential equation $y'' + s(x)y = 0$, where $s(x)$ is the piecewise-defined function (sometimes called the *sign function*):

$$s(x) = \begin{cases} 1 & \text{if } x > 0 \\ -1 & \text{if } x < 0 \\ 0 & \text{if } x = 0. \end{cases}$$

   Can you solve this?

3. Find a particular solution to the ODE $y'' - y' + 2y = 4x + 12$.

4. Find a particular solution to the ODE $y'' - y' + y = \sin^2(x)$. (Hint: it may be helpful to use a trigonometric identity.)

5. Find the general solution to $y^{(3)} - y' = e^x$.

   For the following two problems, determine the form of the particular solution—note that you do not have to determine the values of the coefficients. You should not include terms from the homogeneous solution.

   (a) Determine the general form of the particular solution to $y''' = 9x^2 + 1$.

   (b) Determine the general form of the particular solution to $y^{(4)} - 16y'' = x^2\sin(4x) + \sin(4x)$.

6. Solve the initial value problem $y'' + 2y' + 2y = \cos(3x)$, $y(0) = 0$, $y'(0) = 2$.

7. Solve the initial value problem $y^{(4)} - y = 1$, $y(0) = y'(0) = y''(0) = y^{(3)} = 0$.

2.6 Variation of parameters

The *method of variation of parameters* is originally attributed to Joseph Louis Lagrange (1736 - 1813), an Italian-born mathematician and astronomer, who worked much of his professional life in Berlin and Paris [L-var]. It involves, at one step, a repeated differentiation which is theoretically described by the so-called Leibniz rule, described next.

2.6.1 The Leibniz rule

In general, the *Leibniz rule* (or generalized product rule) is

$$\begin{aligned}(fg)' &= f'g + fg',\\ (fg)'' &= f''g + 2f'g' + fg'',\\ (fg)''' &= f'''g + 3f''g' + 3f'g'' + fg''',\end{aligned}$$

and so on, where the coefficients occurring in $(fg)^{(n)}$ are *binomial coefficients* $\binom{n}{i}$ computed, for example,[4] using *Pascal's triangle*,

$$\begin{array}{ccccccccc} & & & & 1 & & & & \\ & & & 1 & & 1 & & & \\ & & 1 & & 2 & & 1 & & \\ & 1 & & 3 & & 3 & & 1 & \\ 1 & & 4 & & 6 & & 4 & & 1 \\ & & & & \vdots, & & & & \end{array}$$

and so on. The general pattern is dictated by the rule

$$\begin{array}{ccc} \dots \binom{n-1}{i} & \binom{n-1}{i+1} \dots \\ & \dots \binom{n}{i} \quad \dots . \end{array}$$

Using Sage, Leibniz's rule can be checked using the commands below.

Sage

```
sage: t = var('t')
sage: x = function('x', t)
sage: y = function('y', t)
sage: diff(x(t)*y(t),t)
```

[4] For second-order ODEs, we need only the first two Leibniz rules.

```
x(t)*diff(y(t), t, 1) + y(t)*diff(x(t), t, 1)
sage: diff(x(t)*y(t),t,t)
x(t)*diff(y(t), t, 2) + 2*diff(x(t), t, 1)*diff(y(t), t, 1)
 + y(t)*diff(x(t), t, 2)
sage: diff(x(t)*y(t),t,t,t)
x(t)*diff(y(t), t, 3) + 3*diff(x(t), t, 1)*diff(y(t), t, 2)
 + 3*diff(x(t), t, 2)*diff(y(t), t, 1) + y(t)*diff(x(t), t, 3)
```

2.6.2 The method

Consider an ordinary constant-coefficient nonhomogeneous second-order linear differential equation,

$$ay'' + by' + cy = f(x)$$

where $f(x)$ is a given function and a, b, and c are constants. (The varation of parameters method works even if a, b, and c depend on the independent variable x. However, for simplicity, we assume that they are constants here.)

Let $y_1(x)$, $y_2(x)$ be fundamental solutions of the corresponding homogeneous equation

$$ay'' + by' + cy = 0.$$

Start by assuming that there is a particular solution in the form

$$y_p(x) = u_1(x)y_1(x) + u_2(x)y_2(x), \tag{2.11}$$

where $u_1(x)$, $u_2(x)$ are unknown functions [V-var]. We want to solve for u_1 and u_2.

By assumption, y_p solves the ODE, so

$$ay_p'' + by_p' + cy_p = f(x).$$

After some algebra, this becomes

$$a(u_1'y_1 + u_2'y_2)' + a(u_1'y_1' + u_2'y_2') + b(u_1'y_1 + u_2'y_2) = F.$$

If we *assume*

$$u_1'y_1 + u_2'y_2 = 0$$

then we get massive simplification:

$$a(u_1'y_1' + u_2'y_2') = F.$$

Cramer's rule (Lemma 2.3.1) implies that the solution to this system is

$$u_1' = \frac{\det\begin{pmatrix} 0 & y_2 \\ f(x) & y_2' \end{pmatrix}}{\det\begin{pmatrix} y_1 & y_2 \\ y_1' & y_2' \end{pmatrix}}, \qquad u_2' = \frac{\det\begin{pmatrix} y_1 & 0 \\ y_1' & f(x) \end{pmatrix}}{\det\begin{pmatrix} y_1 & y_2 \\ y_1' & y_2' \end{pmatrix}}. \tag{2.12}$$

(Note that the Wronskian $W(y_1, y_2)$ of the fundamental solutions is in the denominator.) Solve these for u_1 and u_2 by integration and then plug them back into (2.11) to get your particular solution.

Example 2.6.1. Solve

$$y'' + y = \tan(x).$$

Solution. The functions $y_1 = \cos(x)$ and $y_2 = \sin(x)$ are fundamental solutions with Wronskian $W(\cos(x), \sin(x)) = 1$. The formulas (2.12) become:

$$u_1' = \frac{\det\begin{pmatrix} 0 & \sin(x) \\ \tan(x) & \cos(x) \end{pmatrix}}{1}, \qquad u_2' = \frac{\det\begin{pmatrix} \cos(x) & 0 \\ -\sin(x) & \tan(x) \end{pmatrix}}{1}.$$

Therefore,

$$u_1' = -\frac{\sin^2(x)}{\cos(x)}, \quad u_2' = \sin(x).$$

Using methods from integral calculus, $u_1 = -\ln|\tan(x) + \sec(x)| + \sin(x)$ and $u_2 = -\cos(x)$. Using **Sage**, this can be checked as follows:

Sage

```
sage: integral(-sin(t)^2/cos(t),t)
-log(sin(t) + 1)/2 + log(sin(t) - 1)/2 + sin(t)
sage: integral(cos(t)-sec(t),t)
sin(t) - log(tan(t) + sec(t))
sage: integral(sin(t),t)
-cos(t)
```

As you can see, there are other forms the answer can take. The particular solution is

$$y_p = (-\ln|\tan(x) + \sec(x)| + \sin(x))\cos(x) + (-\cos(x))\sin(x),$$

which simplifies to

$$y_p = -(\ln|\tan(x) + \sec(x)|)\cos(x).$$

The homogeneous (or complementary) part of the solution is

$$y_h = c_1\cos(x) + c_2\sin(x),$$

so the general solution is

$$\begin{aligned} y &= y_h + y_p = c_1\cos(x) + c_2\sin(x) \\ &\quad -(\ln|\tan(x) + \sec(x)|)\cos(x). \end{aligned}$$

Exercises

1. Find the general solution to $y'' + 4y = \frac{2}{\cos(2x)}$ using variation of parameters.

2. Use the variation of parameters method to find the general solution of

$$y'' - 2y' + y = e^x/x.$$

3. Use the variation of parameters method to solve $x'' + 2x = \tan(2t)$.

4. Use Sage to solve $y'' + y = \cot(x)$.

2.7 Applications of DEs: Spring problems

Ut tensio, sic vis.[5]

—*Robert Hooke, 1678*

2.7.1 Introduction: Simple harmonic case

One of the ways DEs arise is by means of modeling physical phenomenon, such as spring equations. For these problems, consider a spring suspended from a ceiling. We shall consider three cases: (1) no mass is attached at the end of the spring; (2) a mass is attached and the system is in the rest position; (3) a mass is attached and the mass has been displaced from the rest position.

One can also align the springs left to right instead of top to bottom, without changing the discussion below.

Notation. Consider the first two situations above: (a) a spring at rest, without mass attached and (b) a spring at rest, with mass attached. The distance the mass pulls the spring down is sometimes called the "stretch" and denoted s. (A formula for s will be given later.)

Now place the mass in motion by imparting some initial velocity (tap it upward with a hammer, say, and start your timer). Consider the second two situations above: (a) a spring at rest, with mass attached and (b) a spring in motion. The difference between these two

[5] "As the extension, so the force."

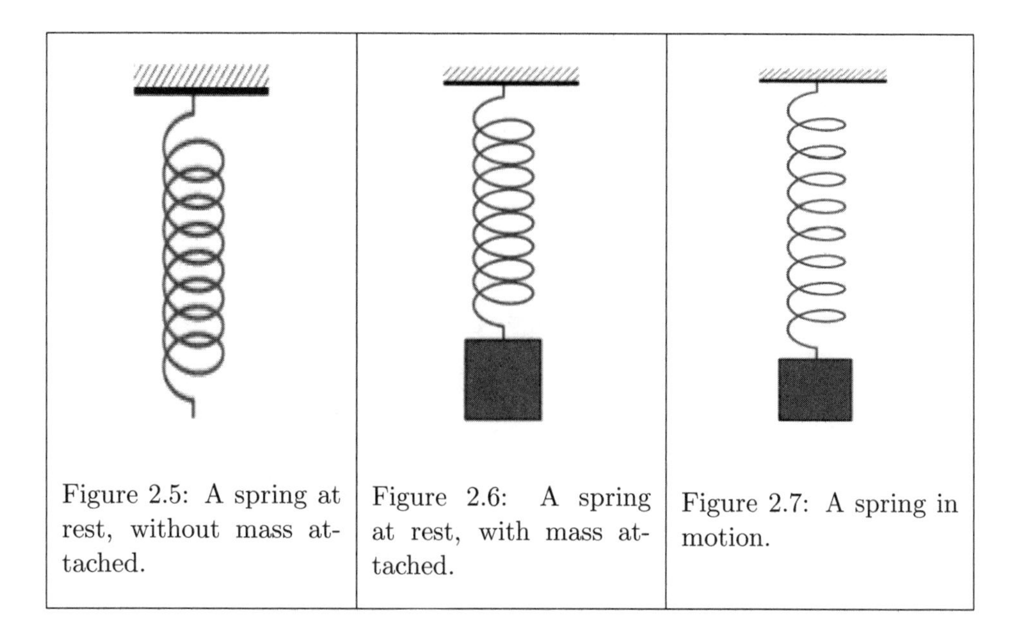

Figure 2.5: A spring at rest, without mass attached.

Figure 2.6: A spring at rest, with mass attached.

Figure 2.7: A spring in motion.

positions at time t is called the *displacement* and is denoted $x(t)$. Signs here will be chosen so that down is positive.

Assume exactly three forces act:

1. the restoring force of the spring, F_{spring};
2. an external force (driving the ceiling up and down, but may be 0), F_{ext};
3. a damping force (imagining the spring immersed in oil or that it is in fact a shock absorber on a car), F_{damp}.

In other words, the total force is given by

$$F_{total} = F_{spring} + F_{ext} + F_{damp}.$$

Physics tells us that the following are approximately true:

1. (*Hooke's law* [H-intro]): $F_{spring} = -kx$, for some "spring constant" $k > 0$;
2. $F_{ext} = F(t)$, for some (possibly zero) function F;
3. $F_{damp} = -bv$, for some "damping constant" $b \geq 0$ (where v denotes velocity);
4. (*Newton's second law* [N-mech]): $F_{total} = ma$ (where a denotes acceleration).

Putting this all together, we obtain $mx'' = ma = -kx + F(t) - bv = -kx + F(t) - bx'$, or

$$\boxed{mx'' + bx' + kx = F(t).} \tag{2.13}$$

This is the *spring equation*.

2.7.2 Simple harmonic case

When $b = F(t) = 0$, (2.13) is also called the equation for *simple harmonic* motion. In this case, sometimes the differential equation is written in the form

$$x'' + \omega^2 x = 0,$$

where $\omega = \sqrt{k/m}$. The solution in the case of simple harmonic motion has the form

$$x(t) = c_1 \cos(\omega t) + c_2 \sin(\omega t),$$

where c_1, c_2 are constants depending on the initial conditions (if any). The graph of this function repeats in "cycles" of $2\pi/\omega$, i.e., it is *periodic* with period $P = 2\pi/\omega$. We call the number $1/P$ the *frequency* of the spring's motion. When initial conditions are given (they usually are), this solution has a more practical form:

$$x(t) = x(0)\cos(\omega t) + \frac{x'(0)}{\omega}\sin(\omega t).$$

There is a more compact form of the solution,

$$x(t) = A\sin(\omega t + \phi),$$

useful for graphing. This compact form is obtained using the general trigonometric identity

$$c_1 \cos(\omega t) + c_2 \sin(\omega t) = A\sin(\omega t + \phi), \qquad (2.14)$$

where $A = \sqrt{c_1^2 + c_2^2}$ denotes the *amplitude* and $\phi = 2\arctan(\frac{c_1}{c_2 + A})$ the *phase shift.*

Consider again Figures 2.5 and 2.6: (a) a spring at rest, without mass attached and (b) a spring at rest, with mass attached. The mass in the second figure is at rest, so the gravitational force on the mass, mg, is balanced by the restoring force of the spring: $mg = ks$, where s is the stretch. In particular, the spring constant can be computed from the stretch:

$$\boxed{k = \frac{mg}{s}.}$$

Example 2.7.1. A spring at rest is suspended from the ceiling without mass. A 2 kg mass is then attached to this spring, stretching it 9.8 cm. From a position 2/3 m above equilibrium the mass is given a downward velocity of 5 m/s.

(a) Find the equation of motion.

(b) What is the amplitude and period of motion?

(c) At what time does the mass first cross equilibrium?

(d) At what time is the mass first exactly 1/2 m below equilibrium?

We shall solve this problem using **Sage** below. Note that $m = 2$, $b = F(t) = 0$ (since no damping or external force is even mentioned), and the stretch is $s = 9.8$ cm $= 0.098$ m. Therefore, the spring constant is given by $k = mg/s = 2 \cdot 9.8/(0.098) = 200$. Therefore, the DE is $2x'' + 200x = 0$. This has general solution $x(t) = c_1 \cos(10t) + c_2 \sin(10t)$. The constants c_1 and c_2 can be computed from the initial conditions $x(0) = -2/3$ (down is positive, up is negative), $x'(0) = 5$.

Using **Sage**, the displacement can be computed as follows:

Sage

```
sage: t = var('t')
sage: x = function('x', t)
sage: m = var('m')
sage: b = var('b')
sage: k = var('k')
sage: F = var('F')
sage: de = lambda y: m*diff(y,t,t) + b*diff(y,t) + k*y - F
sage: de(x)
b*D[0](x)(t) + k*x(t) + m*D[0, 0](x)(t) - F
sage: m = 2; b = 0; k = 200; F = 0
sage: desolve(de(x),[x,t])
k1*sin(10*t)+k2*cos(10*t)
sage: desolve(de(x),[x,t],ics=[0,-2/3,5])
1/2*sin(10*t) - 2/3*cos(10*t)
```

Now we write this in the more compact and useful form $A\sin(\omega t + \phi)$ using formula (2.14) and **Sage**.

Sage

```
sage: c1 = -2/3; c2 = 1/2
sage: A = sqrt(c1^2 + c2^2)
sage: A
5/6
sage: phi = 2*atan(c1/(c2 + A))
sage: phi
-2*atan(1/2)
sage: RR(phi)
-0.927295218001612
sage: sol = lambda t: c1*cos(10*t) + c2*sin(10*t)
sage: sol2 = lambda t: A*sin(10*t + phi)
sage: P = plot(sol(t),0,2)
sage: show(P)
```

This plot is displayed in Figure 2.8.

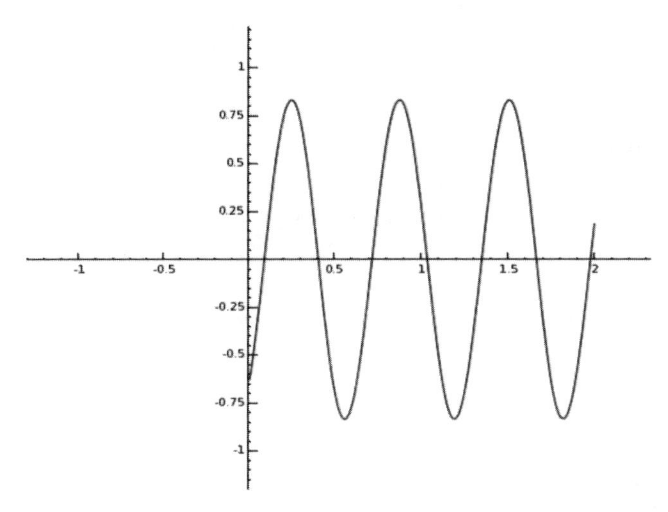

Figure 2.8: Plot of $2x'' + 200x = 0$, $x(0) = -2/3$, $x'(0) = 5$, for $0 < t < 2$.

(You can also, if you want, type `show(plot(sol2(t),0,2))` to check that these two functions are indeed the same.) Of course, the period is $2\pi/10 = \pi/5 \approx 0.628$.

To answer (c) and (d), we solve $x(t) = 0$ and $x(t) = 1/2$:

Sage

```
sage: solve(A*sin(10*t + phi) == 0,t)
[t == atan(1/2)/5]
sage: RR(atan(1/2)/5)
0.0927295218001612
sage: solve(A*sin(10*t + phi) == 1/2,t)
[t == (asin(3/5) + 2*atan(1/2))/10]
sage: RR((asin(3/5) + 2*atan(1/2))/10)
0.157079632679490
```

In other words, $x(0.0927\ldots) \approx 0$, $x(0.157\ldots) \approx 1/2$.

Exercises

1. Using Sage and the problem in Example 2.7.1, answer the following questions.

   (a) At what time does the mass pass through the equilibrium position heading down for the second time?

   (b) At what time is the weight exactly 5/12 m below equilibrium and heading up?

   (c) How many complete vibrations will the mass have completed after 2012 seconds?

2. A spring is suspended from the ceiling. The spring was stretched .098 m after a 1 kg mass was attached to it and has negligable damping. Take $g = 9.8$.

   (a) Find the displacement if the mass is initially released with a 1 m/s downward velocity 2 m up from equilibrium. (Assume "up is positive.")

   (b) Write the displacement in the form $A\sin(\omega t + \phi)$.

   (c) Find the first local extremum of $x(t)$, $t > 0$, and indicate if it is a maximum or a minimum of $x(t)$.

3. A mass of 2 kg is attached to a spring having spring constant 98 N/m with negligible damping. The mass starts at 1 m above equilibrium with initial velocity of 7 m/s downward.

   (a) Find the displacement $x(t)$.

   (b) Write it in the form $A\sin(\omega t + \phi)$.

   (c) Find the second time it crosses equilibrium.

2.7.3 Free damped motion

Recall from §2.7.1 that the spring equation is

$$mx'' + bx' + kx = F(t)$$

where $x(t)$ denotes the displacement at time t.

Until otherwise stated, we assume there is no external force: $F(t) = 0$.

The roots of the characteristic polynomial $mD^2 + bD + k = 0$ are

$$\frac{-b \pm \sqrt{b^2 - 4mk}}{2m},$$

by the quadratic formula. There are three cases:

(a) Real distinct roots: in this case the discriminant $b^2 - 4mk$ is positive, so $b^2 > 4mk$. In other words, b is "large." This case is referred to as *overdamped*. In this case, the roots are negative,

$$r_1 = \frac{-b - \sqrt{b^2 - 4mk}}{2m} < 0 \quad \text{and} \quad r_2 = \frac{-b + \sqrt{b^2 - 4mk}}{2m} < 0,$$

so the solution $x(t) = c_1 e^{r_1 t} + c_2 e^{r_2 t}$ is exponentially decreasing.

(b) Real repeated roots: in this case the discriminant $b^2 - 4mk$ is zero, so $b = \sqrt{4mk}$. This case is referred to as *critically damped*. In this case, the (repeated) roots are negative,

$$r = r_1 = r_2 = -b/2m,$$

so the solution $x(t) = c_1 e^{rt} + c_2 t e^{rt}$ is exponentially decreasing.

If you press down with all your weight on the hood of a new car and then release very quickly, the resulting "bounce" is said to behave as in this case, In other words, new car suspension systems should be modeled by this case [D-spr].

(c) Complex roots: in this case the discriminant $b^2 - 4mk$ is negative, so $b^2 < 4mk$. In other words, b is "small." This case is referred to as *underdamped* (or *simple harmonic* when $b = 0$). The roots are of the form

$$r_1 = \frac{-b - \sqrt{4mk - b^2}i}{2m} \quad \text{and} \quad r_2 = -\frac{b}{2m} + i\frac{\sqrt{4mk - b^2}}{2m} = \alpha + i\beta,$$

so the solution $x(t) = c_1 e^{\alpha t}\cos(\beta t) + c_2 e^{\alpha t}\sin(\beta t)$ is oscillating and $\alpha = -\frac{b}{2m} \leq 0$. When there is damping, this is also exponentially decreasing, sometimes called a "damped oscillation."

Example 2.7.2. An 8 lb weight stretches a spring 2 ft. Assume a damping force numerically equal to twice the instantaneous velocity. Find the displacement at time t, provided that the weight is released from the equilibrium position with an upward velocity of 3 ft/s. Find the equation of motion and classify the behavior.

We know $m = 8/32 = 1/4$, $b = 2$, $k = mg/s = 8/2 = 4$, $x(0) = 0$, and $x'(0) = -3$. This means we must solve

$$\frac{1}{4}x'' + 2x' + 4x = 0, \quad x(0) = 0, \quad x'(0) = -3.$$

The roots of the characteristic polynomial are -4 and -4 (so we are in the repeated real roots case), so the general solution is $x(t) = c_1 e^{-4t} + c_2 t e^{-4t}$. The initial conditions imply $c_1 = 0$, $c_2 = -3$, so

$$x(t) = -3te^{-4t}.$$

Using **Sage**, we can compute this as well:

Sage

```
sage: t = var('t')
sage: x = function('x',t)
sage: de = lambda y: (1/4)*diff(y,t,t) + 2*diff(y,t) + 4*y
```

```
sage: de(x)
4*x(t) + 2*D[0](x)(t) + 1/4*D[0, 0](x)(t)
sage: desolve(de(x),[x,t])
(k2*t + k1)*e^(-4*t)
sage: desolve(de(x),[x,t],ics=[0,0,-3])
-3*t*e^(-4*t)
sage: f = lambda t : -3*t*e^(-4*t)
sage: P = plot(f,0,2)
sage: show(P)
```

The graph is shown in Figure 2.9.

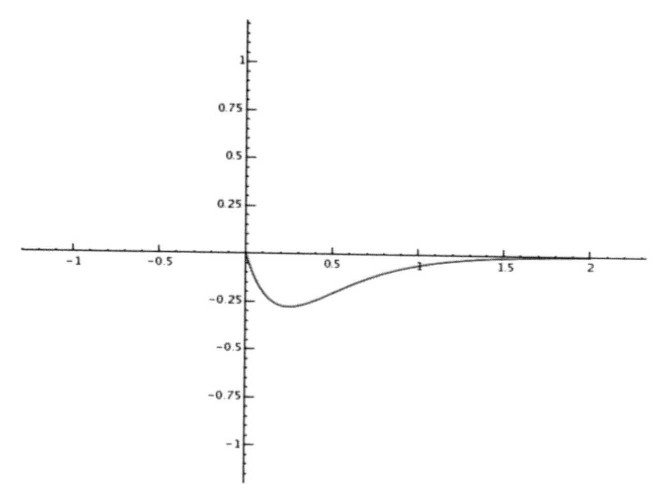

Figure 2.9: Plot of $(1/4)x'' + 2x' + 4x = 0$, $x(0) = 0$, $x'(0) = -3$, for $0 < t < 2$.

Example 2.7.3. A 2 kg mass is attached to a spring having spring constant 10. Assume a damping force numerically equal to four times the instantaneous velocity. Find the displacement at time t, provided that the mass is released from 1 m below equilibrium with an upward velocity of 1 ft/s. Find the equation of motion and classify the behavior.

Using Sage, we can compute this as well:

Sage

```
sage: t = var('t')
sage: x = function('x',t)
sage: de = lambda y: 2*diff(y,t,t) + 4*diff(y,t) + 10*y
sage: desolve(de(x),[x,t],ics=[0,1,1])
(sin(2*t) + cos(2*t))*e^(-t)
sage: desolve(de(x),[x,t],ics=[0,1,-1])
e^(-t)*cos(2*t)
```

```
sage: sol = lambda t: e^(-t)*cos(2*t)
sage: P = plot(sol(t),0,2)
sage: show(P)
sage: P = plot(sol(t),0,4)
sage: show(P)
```

The graph is shown in Figure 2.10.

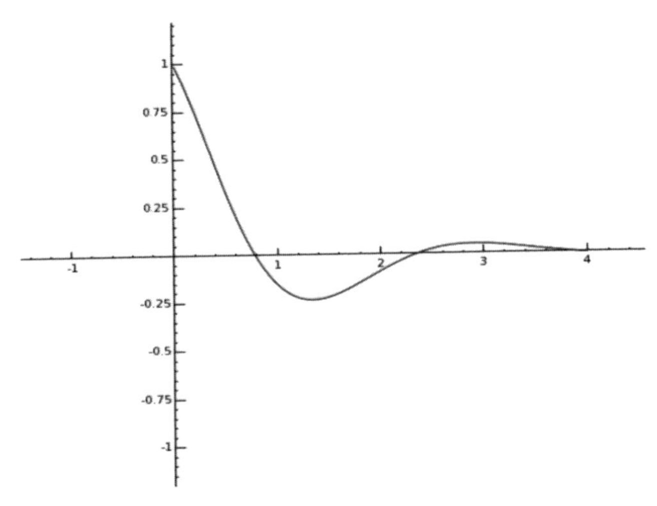

Figure 2.10: Plot of $2x'' + 4x' + 10x = 0$, $x(0) = 1$, $x'(0) = -1$, for $0 < t < 4$.

Exercises

1. By hand, solve $x'' + 5x' + 6x = 0$, $x(0) = 1$, $x'(0) = 2$. Find the velocity at $t = 1$.

2. Refer to Example 2.7.3 above. Use Sage to find at what time the weight passes through the equilibrium position heading down for the second time.

3. A 2 kg mass is attached to a spring having spring constant 10. Assume a damping force numerically equal to four times the instantaneous velocity. Use Sage to find the displacement at time t, provided that the mass is released from 1 m below equilibrium (with no initial velocity).

4. A 2 lb weight is attached to a spring and stretches the spring 1/5 ft. An additional 2 lb weight is attached to the spring, and after this the spring is released from rest at a point 2 ft above its equilibrium position. The damping force is equal to the instantaneous velocity in absolute value. No external driving force is applied to this spring-weight system.

(a) Write the differential equation and initial conditions that model the spring motion in terms of the displacement $x(t)$ and time t.

(b) Solve this differential equation.

(c) State whether the spring is overdamped, underdamped, or critically damped.

5. A 2 kg mass is attached to a spring having spring constant $k = 26$, with damping constant $b = 12$. The mass is initially released with a 2 m/s downward velocity 1 m above equilibrium. Find the displacement at time t.

2.7.4 Spring-mass systems with an external force

If the frequency of the driving force of the spring matches the frequency of the homogeneous part $x_h(t)$, in other words if

$$x'' + \omega^2 x = F_0 \cos(\gamma t)$$

satisfies $\omega = \gamma$ then we say that the spring-mass system is in *(pure, mechanical) resonance.* For some time, it was believed that the collapse of the Tacoma Narrows Bridge was explained by this phenomenon, but this is now known to be false.

Theorem 2.7.1. Consider

$$x'' + \omega^2 x = F_0 \cos(\gamma t) \tag{2.15}$$

and

$$x'' + \omega^2 x = F_0 \sin(\gamma t). \tag{2.16}$$

Let $x_p = x_p(t)$ denote a particular solution to either (2.15) or (2.16). Then we have the following cases.

- Case (2.15):

  1. if $\gamma \neq \omega$ then

$$x_p(t) = \frac{F_0}{\omega^2 - \gamma^2} \cos(\gamma t)$$

  2. if $\gamma = \omega$ then

$$x_p(t) = \frac{F_0}{2\omega} t \sin(\gamma t)$$

- Case (2.16):

1. if $\gamma \neq \omega$ then

$$x_p(t) = \frac{F_0}{\omega^2 - \gamma^2}\sin(\gamma t)$$

2. if $\gamma = \omega$ then

$$x_p(t) = -\frac{F_0}{2\omega} t\cos(\gamma t).$$

In particular, if $\gamma \neq \omega$ then

$$x(t) = \frac{F_0}{\omega^2 - \gamma^2}(\cos(\gamma t) - \cos(\omega t))$$

is a solution to (2.15). Likewise, if $\gamma \neq \omega$ then

$$x(t) = \frac{F_0}{\omega^2 - \gamma^2}(\sin(\gamma t) - \sin(\omega t))$$

is a solution to (2.16). In both of these, to derive the case $\gamma = \omega$, one can take limits $\gamma \to \omega$ in these expressions. Use L'Hôpital's rule to compute the limits. (Details are left to the interested reader.)

Example 2.7.4. Solve

$$x'' + \omega^2 x = \cos(\gamma t), \quad x(0) = 0, \quad x'(0) = 0,$$

where $\omega = \gamma = 2$ (i.e., mechanical resonance). We use Sage for this:

Sage

```
sage: t = var('t')
sage: x = function('x', t)
sage: (m,b,k,w,F0) = var("m,b,k,w,F0")
sage: de = lambda y: diff(y,t,t) + w^2*y - F0*cos(w*t)
sage: m = 1; b = 0; k = 4; F0 = 1; w = 2
sage: desolve(de(x),[x,t])
k1*sin(2*t) + k2*cos(2*t) + 1/4*t*sin(2*t) + 1/8*cos(2*t)
sage: soln = lambda t : t*sin(2*t)/4   # this is the soln satisfying the ICs
sage: P = plot(soln(t),0,10)
sage: show(P)
```

This is displayed in Figure 2.11.

Example 2.7.5. Solve

$$x'' + \omega^2 x = \cos(\gamma t), \quad x(0) = 0, \quad x'(0) = 0,$$

where $\omega = 2$ and $\gamma = 3$ (i.e., no mechanical resonance). We use Sage for this:

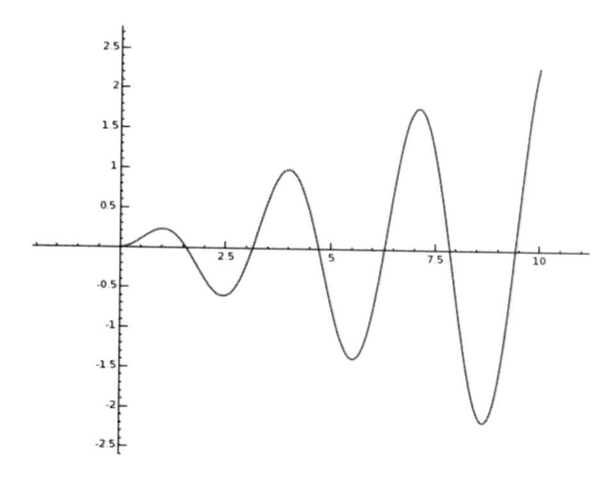

Figure 2.11: A forced undamped spring, with resonance.

Sage

```
sage: t = var('t')
sage: x = function('x', t)
sage: (m,b,k,w,g,F0) = var("m,b,k,w,g,F0")
sage: de = lambda y: diff(y,t,t) + w^2*y - F0*cos(g*t)
sage: m = 1; b = 0; k = 4; F0 = 1; w = 2; g = 3
sage: desolve(de(x),[x,t])
k1*sin(2*t) + k2*cos(2*t) - 1/5*cos(3*t)
desolve(de(x),[x,t],[0,0,0])
1/5*cos(2*t) - 1/5*cos(3*t)
sage: soln = lambda t : cos(2*t)/5-cos(3*t)/5
sage: P = plot(soln(t),0,10)
sage: show(P)
```

This is displayed in Figure 2.12.

Exercises

1. A mass-spring system with external force is governed by the differential equation $4x'' + 16x = 5\cos\omega t$. The resonant angular frequency ω of the system is

   (a) 0, (b) 1, (c) 2, (d) 4.

2. An 8 lb weight stretches a spring 2 ft upon coming to rest at equilibrium. From equilibrium the weight is raised 1 ft and released from rest. The motion of the weight is resisted by a damping force that in numerically equal to twice the weight's instantaneous velocity.

   (a) Find the position of the weight as a function of time.

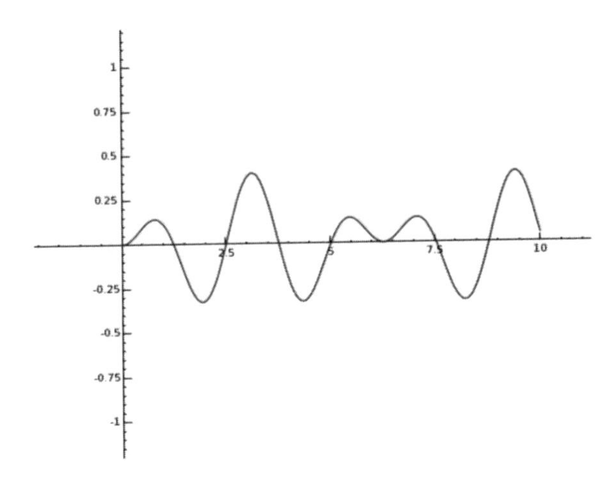

Figure 2.12: A forced undamped spring, no resonance.

(b) What type of damping does this mass-spring system possess?

3. The position of a weight in a mass-spring system subject to an external force is given by $x(t) = e^{-t}\cos(3t) + e^{-t}\sin(3t) + 6\cos(2t) + 4\sin(2t)$.

   (a) What are the amplitude and period of the steady-state part of the solution?

   (b) Write the transient part of the solution in the form $Ae^{-t}\sin(3t+\phi)$.

   (c) Find the time after which the magnitude of the transient part of the solution is less than 1% of that of the steady-state part of the solution.

4. If the displacement $x = x(t)$ for an object attached to a spring suspended from a ceiling is governed by the differential equation $2x'' + 6x' + 4x = 0$ then its behavior is
   (a) in mechanical resonance, (b) underdamped, (c) critically damped,
   (d) overdamped, (e) none of these.

5. The amplitude of the steady-state oscillations of a damped mass-spring system with a sinusoidal external force $F = \cos\omega t$ is given by $C(\omega) = \frac{1}{\sqrt{(\omega^2-4)^2+4\omega^2}}$. The resonant angular frequency ω of this system is
   (a) 0, (b) $\frac{1}{\sqrt{3}}$, (c) $\sqrt{2}$, (d) $\sqrt{3}$.

6. A spring-mass system is governed by $2x'' + 50x = 10\sin(\omega t)$. The system has mechanical resonance when ω equals
   (a) 7, (b) $\sqrt{50}$, (c) 1, (d) 5, (e) none of these.

7. (a) A 4 lb weight is suspended from a spring with spring constant $k = 7/8$ lb/ft and damping constant $b = 1/2$ lb-s/ft. If the weight is pulled 1 ft below equilibrium and given a velocity of 5 ft/s upward, find its displacement in the form $x(t) = \exp(\alpha t) \cdot A\sin(\omega t + \phi)$.

(b) What value of $b > 0$ would give critical damping? If $b = 0$ and there is an external force of $10\cos(t)$, what value of b would give resonance?

- (*) Verify all the cases of Theorem 2.7.1 stated above using the method of undetermined coefficients.
- Complete the proof of Theorem 2.7.1 as sketched in the text.

2.8 Applications to simple LRC circuits

An LRC circuit is a closed loop containing an inductor of L henries, a resistor of R ohms, a capacitor of C farads, and an emf (electromotive force), or battery, of $E(t)$ volts, all connected in series.

These circuits arise in several engineering applications. For example, AM/FM radios with analog tuners typically use an LRC circuit to tune a radio frequency. Most commonly a variable capacitor is attached to the tuning knob, which allows you to change the value of C in the circuit and tune to stations on different frequencies [R-cir].

We use the "dictionary" in Figure 2.13 to translate between the electrical engineering (EE) diagram and the differential equations.

EE object	term in DE (the voltage drop)	units	symbol
charge	$q = \int i(t)\,dt$	coulombs C	
current	$i = q'$	amps A	
emf	$e = e(t)$	volts V	
resistor	$Rq' = Ri$	ohms Ω	
capacitor	$C^{-1}q$	farads F	
inductor	$Lq'' = Li'$	henries H	

Figure 2.13: Dictionary for electrical circuits.

Next, we recall the circuit laws of Gustav Kirchoff (sometimes spelled Kirchhoff), a German physicist who lived from 1824 to 1887. He was born in Königsberg, which was part of Germany but is now part of the Kaliningrad Oblast, an exclave of Russia surrounded by Lithuania, Poland, and the Baltic Sea.

Kirchoff's first law: The algebraic sum of the currents traveling into any node is zero.

Kirchoff's second law: The algebraic sum of the voltage drops around any closed loop is zero.

Generally, the charge at time t on the capacitor, $q(t)$, satisfies the differential equation

$$Lq'' + Rq' + \frac{1}{C}q = E(t). \tag{2.17}$$

When there is no electromotive force, that is, when

$$Lq'' + Rq' + \frac{1}{C}q = 0,$$

then we say that the circuit is

- *underdamped* if $R^2 - 4L/C < 0$,
- *critically damped* if $R^2 - 4L/C = 0$, and
- *overdamped* if $R^2 - 4L/C > 0$.

Example 2.8.1. In this example, we model a very simple type of radio tuner, using a variable capacitor to represent the tuning dial. Consider the simple LC circuit given by the diagram in Figure 2.14.

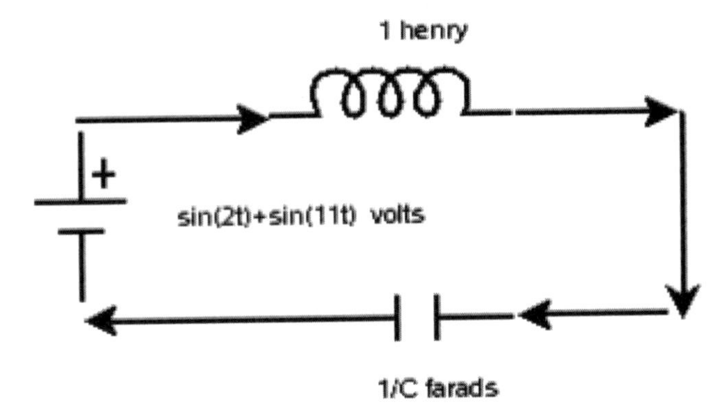

Figure 2.14: A simple LC circuit.

According to Kirchoff's second law and the above "dictionary," this circuit corresponds to the differential equation

$$q'' + \frac{1}{C}q = \sin(2t) + \sin(11t).$$

The homogeneous part of the solution is

$$q_h(t) = c_1 \cos(t/\sqrt{C}) + c_1 \sin(t/\sqrt{C}).$$

If $C \neq 1/4$ and $C \neq 1/121$ then

$$q_p(t) = \frac{1}{C^{-1} - 4}\sin(2t) + \frac{1}{C^{-1} - 121}\sin(11t).$$

When $C = 1/4$ and the initial charge and current are both zero, the solution is

$$q(t) = -\frac{1}{117}\sin(11t) + \frac{161}{936}\sin(2t) - \frac{1}{4}t\cos(2t).$$

This can be computed using Sage as well:

Sage

```
sage: t = var("t")
sage: q = function("q",t)
sage: L,R,C = var("L,R,C")
sage: E = lambda t:sin(2*t)+sin(11*t)
sage: de = lambda y: L*diff(y,t,t) + R*diff(y,t) + (1/C)*y-E(t)
sage: L,R,C=1,0,1/4
sage: de(q)
-sin(2*t) - sin(11*t) + 4*q(t) + D[0, 0](q)(t)
sage: desolve(de(q(t)),[q,t],[0,0,0])
-1/4*t*cos(2*t) + 161/936*sin(2*t) - 1/117*sin(11*t)
sage: soln = lambda t: -sin(11*t)/117+161*sin(2*t)/936-t*cos(2*t)/4
sage: P = plot(soln,0,10)
sage: show(P)
```

This is displayed in Figure 2.15.

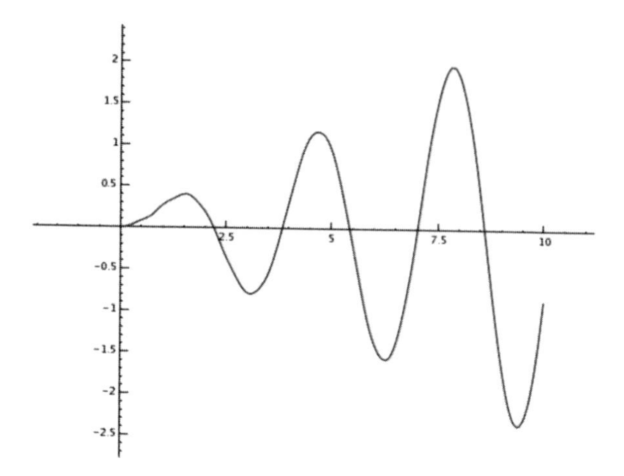

Figure 2.15: An LC circuit, with resonance.

You can see how the frequency $\omega = 2$ dominates the other terms.

When $0 < R < 2\sqrt{L/C}$ the homogeneous form of the charge in (2.17) has the form

$$q_h(t) = c_1 e^{\alpha t}\cos(\beta t) + c_2 e^{\alpha t}\sin(\beta t),$$

where $\alpha = -R/2L < 0$ and $\beta = \sqrt{4L/C - R^2}/(2L)$. This is sometimes called the *transient part* of the solution. The remaining terms in the charge (those corresponding to the "particular" q_p, when using the method of undertermined coefficients to solve the ODE) are called the *steady-state terms.*

Example 2.8.2. An LRC circuit has a 1 H inductor, a 2 Ω resistor, a 1/5 F capacitor, and an emf of $50\cos(t)$ V. If the initial charge and current are 0, find the charge at time t.

The initial value problem describing the charge $q(t)$ is

$$q'' + 2q' + 5q = 50\cos(t), \quad q(0) = q'(0) = 0.$$

The homogeneous part of the solution is

$$q_h(t) = c_1 e^{-t}\cos(2t) + c_2 e^{-t}\sin(2t).$$

The general form of the particular solution using the method of undetermined coefficients is

$$q_p(t) = A_1\cos(t) + A_2\sin(t).$$

Solving for A_1 and A_2 gives $A_1 = 10$, $A_2 = 5$, so the initial conditions give

$$q(t) = q_h(t) + q_p(t) = -10e^{-t}\cos(2t) - \frac{15}{2}e^{-t}\sin(2t) + 10\cos(t) + 5\sin(t).$$

The term

$$q_p(t) = 10\cos(t) + 5\sin(t)$$

is sometimes called the *steady state charge.*

We can also solve this using Sage.

Sage

```
sage: t = var("t")
sage: q = function("q",t)
sage: L,R,C = var("L,R,C")
sage: E = lambda t: 50*cos(t)
sage: de = lambda y: L*diff(y,t,t) + R*diff(y,t) + (1/C)*y-E(t)
sage: L,R,C = 1,2,1/5
sage: desolve(de(q),[q,t])
(k1*sin(2*t) + k2*cos(2*t))*e^(-t) + 5*sin(t) + 10*cos(t)
sage: desolve(de(q),[q,t], [0,0,0])
-5/2*(3*sin(2*t) + 4*cos(2*t))*e^(-t) + 5*sin(t) + 10*cos(t)
sage: soln = lambda t: e^(-t)*(-15*sin(2*t)/2-10*cos(2*t))+5*sin(t)+10*cos(t)
sage: P = plot(soln,0,10)
sage: soln_ss = lambda t: 5*sin(t)+10*cos(t)
sage: P_ss = plot(soln_ss,0,10,linestyle=":")
```

```
sage: soln_tr = lambda t: e^(-t)*(-15*sin(2*t)/2-10*cos(2*t))
sage: P_tr = plot(soln_tr,0,10,linestyle="--")
sage: show(P+P_ss+P_tr)
```

This plot (the solution superimposed with the transient part of the solution) is displayed in Figure 2.16.

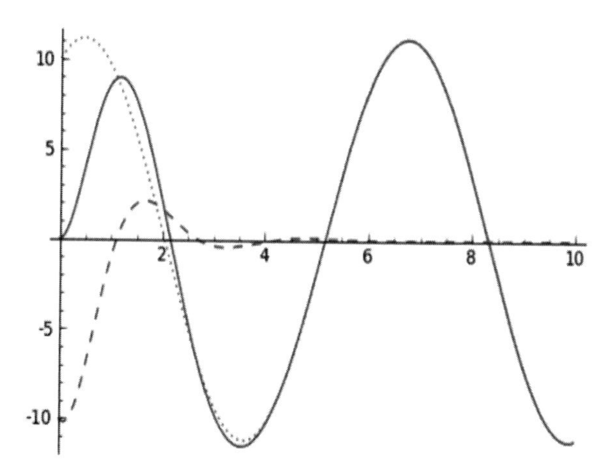

Figure 2.16: An LRC circuit, with damping, and the transient part (dashed) of the solution.

Exercises

1. Use Sage to solve

$$q'' + \frac{1}{C}q = \sin(2t) + \sin(11t), \quad q(0) = q'(0) = 0,$$

in the cases $C = 1/121$, $C = 1/121 + 10^{-4}$, and $C = 1$. Plot the graph of the three charges, for $0 \le t \le 2\pi$, superimposed on the same plot, but with different colors.

2. A simple circuit has an 8 Ω resistor and a 1/17 F capacitor. A battery of $\sin(t)$ V is attached. Assume that the initial charge is zero. Find the charge at time t.

3. A circuit has an emf of 20 V, an inductor of 4 H, and a capacitor of .01 F. If the initial current is 5 A and initial charge on the capacitor is 1.2 C, find the charge at time t.

4. For an LRC series circuit with $L = 2$ H, $R = 14$ Ω, $C = 0.05$ F, and an electromotive force of $E = 120$ V, the steady-state charge on the capacitor is $q =$

(a) 120, (b) 6, (c) $60t$, (d) $120t$.

5. An RC circuit has an electromotive force given by $E(t) = 100\sin(100t)$ V, a resistor of 1000 Ω and a capacitor of .0005 F. If the initial charge on the capacitor is zero, then do the following:

   (a) Draw a schematic diagram of the circuit.

   (b) Set up and solve an initial value problem for the charge $q(t)$ at any time.

   (c) If the voltage across the capacitor is given by $v_c(t) = \frac{1}{C}q(t)$, find the steady-state voltage $v(t)$ across the capacitor.

   (d) Express the steady state $v(t)$ found in (c) in the form $A\sin(\omega t + \phi)$ where you must find A and ϕ.

6. An RLC circuit has a 1 H inductor in series with a 4 Ω resistor and a 1/5 F capacitor. A battery of $\sin(t)$ V is attached.

   (a) Assume that the initial charge and current are zero. Find the current at time t.

   (b) Identify the transient and steady state terms.

7. An RLC circuit has a 1/2 H inductor in series with a 3 Ω resistor and a 1/5 F capacitor. A battery of $39\sin(t)$ V is attached.

   (a) Assume that the initial charge and current are zero. Find the charge at time t.

   (b) Identify the transient and steady-state terms.

8. Solve each differential equation and say if it is underdamped, overdamped, or critically damped.

$$\begin{aligned} q'' + 5q' + 4q &= 0, \;\; q(0) = q'(0) = 1, \\ q'' + 2q' + 10q &= 0, \;\; q(0) = -2, \; q'(0) = 0. \end{aligned}$$

2.9 The power series method

2.9.1 Part 1

In this part, we recall some basic facts about power series and Taylor series. We will turn to solving ODEs in Part 2.

Roughly speaking, *power series* are simply infinite degree polynomials

$$f(x) = a_0 + a_1 x + a_2 x^2 + \cdots = \sum_{k=0}^{\infty} a_k x^k, \tag{2.18}$$

for some real or complex numbers $a_0, a_1, \ldots$. A power series is a way of expressing a "complicated" function $f(x)$ as a sum of "simple" functions like x, x^2, ... The number

a_k is called the *coefficient* of x^k, for $k = 0, 1, \ldots$. Let us ignore for the moment the precise meaning of this infinite sum. (How do you associate a value with an infinite sum? Does the sum converge for some values of x? If so, for which values?) We will return to that issue later.

First, some motivation. Why study these? This type of function is convenient for several reasons.

- It is easy to differentiate a power series (term by term):

$$f'(x) = a_1 + 2a_2x + 3a_3x^2 + \cdots = \sum_{k=0}^{\infty} ka_kx^{k-1} = \sum_{k=0}^{\infty}(k+1)a_{k+1}x^k.$$

- It is easy to integrate such a series (term by term):

$$\int f(x)\,dx = a_0x + \frac{1}{2}a_1x^2 + \frac{1}{3}a_2x^3 + \cdots = \sum_{k=0}^{\infty}\frac{1}{k+1}a_kx^{k+1} = \sum_{k=1}^{\infty}\frac{1}{k}a_{k-1}x^k.$$

- If the summands a_kx^k tend to zero very quickly, then the sum of the first few terms of the series is often a good numerical approximation for the function itself.
- Power series enable one to reduce the solution of certain differential equations down to the (often much easier) problem of solving certain recurrence relations.
- Power series expansions arise naturally in Taylor's theorem from differential calculus.

Theorem 2.9.1. (*Taylor's Theorem*) If $f(x)$ is $n+1$ times continuously differentiable in the open interval (a, x) then there exists a point $\xi \in (a, x)$, such that

$$f(x) = f(a) + (x-a)f'(a) + \frac{(x-a)^2}{2!}f''(a) + \cdots + \frac{(x-a)^n}{n!}f^{(n)}(a) + \frac{(x-a)^{n+1}}{(n+1)!}f^{(n+1)}(\xi). \tag{2.19}$$

The sum

$$T_n(x) = f(a) + (x-a)f'(a) + \frac{(x-a)^2}{2!}f''(a) + \cdots + \frac{(x-a)^n}{n!}f^{(n)}(a),$$

is called the *nth degree Taylor polynomial of f centered at a*. For the case $n = 1$, the formula is

$$T_1(x) = f(a) + (x-a)f'(a),$$

which is the tangent-line approximation to $f(x)$ at $x = a$.

The *Taylor series of f centered at a* is the series

$$f(a)+(x-a)f'(a)+\frac{(x-a)^2}{2!}f''(a)+\cdots .$$

When this series converges to $f(x)$ at each point x in some inverval centered about a then we say f has a *Taylor series expansion* (or Taylor series representation) at a. A Taylor series is basically just a power series, but using powers of $x-a$ instead of powers of just x.

As the examples below indicate, many of the functions you are used to seeing from calculus have a Taylor series representation.

- Geometric series:

$$\begin{aligned}\frac{1}{1-x} &= 1+x+x^2+x^3+x^4+\cdots \\ &= \sum_{n=0}^{\infty} x^n \end{aligned} \tag{2.20}$$

To see this, assume $|x|<1$ and let $n\to\infty$ in the polynomial identity

$$1+x+x^2+\cdots+x^{n-1}=\frac{1-x^{n+1}}{1-x}.$$

For $x\geq 1$, the series does not converge.

- The exponential function:

$$\begin{aligned} e^x &= 1+x+\frac{x^2}{2}+\frac{x^3}{6}+\frac{x^4}{24}+\cdots \\ &= 1+x+\frac{x^2}{2!}+\frac{x^3}{3!}+\frac{x^4}{4!}+\cdots \\ &= \sum_{n=0}^{\infty}\frac{x^n}{n!} \end{aligned} \tag{2.21}$$

To see this, take $f(x)=e^x$ and $a=0$ in Taylor's theorem (2.19), using the fact that $\frac{d}{dx}e^x=e^x$ and $e^0=1$:

$$e^x=1+x+\frac{x^2}{2!}+\frac{x^3}{3!}+\cdots+\frac{x^n}{n!}+\frac{\xi^{n+1}}{(n+1)!},$$

for some ξ between 0 and x. Perhaps it is not clear to everyone that as n becomes larger and larger (x fixed), the last ("remainder") term in this sum goes to 0. However, *Stirling's formula* tells us how large the factorial function grows,

$$n!\sim\sqrt{2\pi n}\left(\frac{n}{e}\right)^n\left(1+O(n^{-1})\right),$$

so we may indeed take the limit as $n \to \infty$ to get (2.21).

`Wikipedia`'s entry on "Power series" [P1-ps] has a nice animation showing how more and more terms in the Taylor polynomials approximate e^x better and better. This animation can also be constructed using Sage (see, for example, the interactive feature at
`http://wiki.sagemath.org/interact/calculus#TaylorSeries`).

- The cosine function:

$$\begin{aligned}\cos x &= 1 - \frac{x^2}{2} + \frac{x^4}{24} - \frac{x^6}{720} + \cdots \\ &= 1 - \frac{x^2}{2!} + \frac{x^4}{4!} - \frac{x^6}{6!} + \cdots \\ &= \sum_{n=0}^{\infty} (-1)^n \frac{x^{2n}}{(2n)!}. \end{aligned} \tag{2.22}$$

This too follows from Taylor's theorem (take $f(x) = \cos x$ and $a = 0$).

Sage

```
sage: x = var("x")
sage: cos(x).series(x, 10)
1 + (-1/2)*x^2 + 1/24*x^4 + (-1/720)*x^6 + 1/40320*x^8 + Order(x^10)
```

However, there is another trick to derive this series expansion: Replace x in (2.21) by ix and use the fact ("Euler's formula") that

$$e^{ix} = \cos(x) + i\sin(x).$$

Taking real parts gives (2.22). Taking imaginary parts gives (2.23).

- The sine function:

$$\begin{aligned}\sin x &= x - \frac{x^3}{6} + \frac{x^5}{120} - \frac{x^7}{5040} + \cdots \\ &= 1 - \frac{x^3}{3!} + \frac{x^5}{5!} - \frac{x^7}{7!} + \cdots \\ &= \sum_{n=0}^{\infty} (-1)^n \frac{x^{2n+1}}{(2n+1)!}. \end{aligned} \tag{2.23}$$

Indeed, you can formally check (using formal term by term differentiation) that

$$-\frac{d}{dx}\cos(x) = \sin(x).$$

(Alternatively, you can use this fact to deduce (2.23) from (2.22).)

- The logarithm function:

$$\log(1-x) = -x - \frac{1}{2}x^2 - \frac{1}{3}x^3 - \frac{1}{4}x^4 + \cdots$$
$$= -\sum_{n=0}^{\infty} \frac{1}{n}x^n. \tag{2.24}$$

This follows from (2.20) since (using formal term-by-term integration)

$$\int_0^x \frac{1}{1-t} = -\log(1-x).$$

Sage

```
sage: taylor(sin(x), x, 0, 5)
 x - x^3/6 + x^5/120
sage: P1 = plot(sin(x),0,pi)
sage: P2 = plot(x,0,pi,linestyle="--")
sage: P3 = plot(x-x^3/6,0,pi,linestyle="-.")
sage: P4 = plot(x-x^3/6+x^5/120,0,pi,linestyle=":")
sage: T1 = text("x",(3,2.5))
sage: T2 = text("x-x^3/3!",(3.5,-1))
sage: T3 = text("x-x^3/3!+x^5/5!",(3.7,0.8))
sage: T4 = text("sin(x)",(3.4,0.1))
sage: show(P1+P2+P3+P4+T1+T2+T3+T4)
```

This is displayed in Figure 2.17.

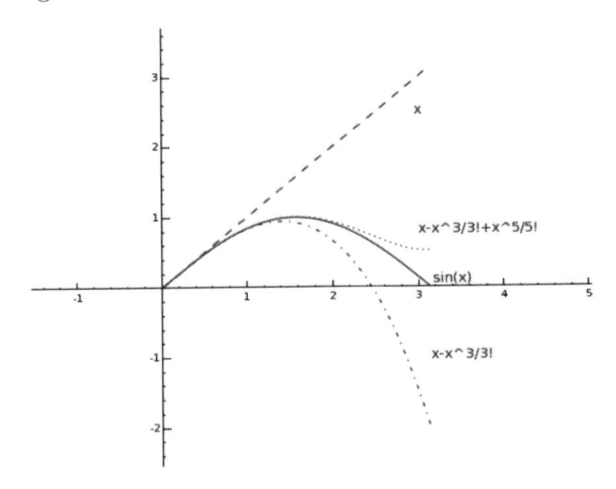

Figure 2.17: Taylor polynomial approximations for $\sin(x)$.

Exercise

Use Sage to plot successive Taylor polynomial approximations for $\cos(x)$.

Finally, we turn to the rigorous meaning of these infinite sums (2.18). How do you associate a value with an infinite sum? Does the sum converge for some values of x? If so, for which values? We will (for the most part) answer all of these questions.

First, consider our infinite power series $f(x)$ in (2.18), where the a_k are all given and x is fixed for the moment. The *partial sums* of this series are

$$f_0(x) = a_0, \quad f_1(x) = a_0 + a_1 x, \quad f_2(x) = a_0 + a_1 x + a_2 x^2, \cdots .$$

We say that the series in (2.18) *converges at x* if the limit of partial sums

$$\lim_{n\to\infty} f_n(x)$$

exists. The following fact is not proven here, as it would take us too far afield. See, for example, Gleason [Gl], Theorem 15-2.2.

Theorem 2.9.2. There is an $R \geq 0$ (possibly ∞) such that the power series (2.18) for $f(x)$ converges absolutely, if $|x| < R$ and diverges if $|x| > R$.

The R in this theorem is called the *radius of convergence.*

There are several tests for determining whether or not a series converges. One of the most commonly used tests is the following theorem.

Theorem 2.9.3. *(Root test)* Assume

$$L = \lim_{k\to\infty} |a_k x^k|^{1/k} = |x| \lim_{k\to\infty} |a_k|^{1/k}$$

exists. If $L < 1$ then the infinite power series $f(x)$ in (2.18) converges at x. In general, (2.18) converges for all x satisfying

$$-\lim_{k\to\infty} |a_k|^{-1/k} < x < \lim_{k\to\infty} |a_k|^{-1/k}.$$

The number $\lim_{k\to\infty} |a_k|^{-1/k}$ (if it exists, though it can be ∞) is called the *radius of convergence.*

Example 2.9.1. The radius of convergence of e^x (and $\cos(x)$ and $\sin(x)$) is ∞. The radius of convergence of $1/(1-x)$ (and $\log(1-x)$) is 1.

Example 2.9.2. The radius of convergence of

$$f(x) = \sum_{k=0}^{\infty} \frac{k^7 + k + 1}{2^k + k^2} x^k$$

can be determined with the help of `Sage`. We want to compute

$$\lim_{k\to\infty} \left| \frac{k^7 + k + 1}{2^k + k^2} \right|^{-1/k}.$$

Sage

```
sage: k = var('k')
sage: limit(((k^7+k+1)/(2^k+k^2))^(-1/k),k=infinity)
2
```

In other words, the series converges for all x satisfying $-2 < x < 2$.

Another useful test for the convergence of a power series is given in the following theorem.

Theorem 2.9.4. *(Ratio test)* Assume

$$L = \lim_{k\to\infty} |\frac{a_{k+1}x^{k+1}}{a_k x^k}| = |x| \lim_{k\to\infty} \frac{|a_{k+1}|}{|a_k|}$$

exists. The ratio test states that:

- if $L < 1$ then the series converges absolutely;
- if $L > 1$ then the series does not converge;
- if $L = 1$ (or if the limit fails to exist) then the test is inconclusive.

Example 2.9.3. The radius of convergence of

$$e^x = \sum_{k=0}^{\infty} \frac{x^k}{k!}$$

is ∞. We have

$$L = |x| \lim_{k\to\infty} \frac{1/(k+1)!}{1/k!} = |x| \lim_{k\to\infty} (k+1)^{-1} = 0,$$

for all x. By the ratio test, e^x converges for all x.

Represention of functions by power series is a powerful technique. You should be aware, however, that there are functions which are infinitely differentiable at a point but not *analytic*—that is, the Taylor series does not converge to the function in any interval containing the point of expansion. One example of such a function is

$$f(t) = \begin{cases} e^{-1/t^2} & \text{if } t \neq 0 \\ 0 & \text{if } t = 0 \end{cases}$$

It can be shown that this function is infinitely differentiable, but all of its derivatives at $t = 0$ are equal to zero. This means its power series expansion around $t = 0$ is the zero function. The power series converges everywhere—to zero—but since $f(t) > 0$ for $t \neq 0$, it is not converging to $f(t)$ in any open interval.

Exercises

1. Use Sage to find the series expansion of $e^{-x}\cos(x)$ to ten terms.

2. Use Sage to find the series expansion of $e^x/(e^x+1)$ to ten terms.

3. Find the radius of convergence of

$$f(x) = \sum_{k=0}^{\infty} \frac{1}{2^k} x^k$$

using the root test.

4. Find the radius of convergence of

$$f(x) = \sum_{k=0}^{\infty} \frac{1}{2^k} (x-1)^k$$

using the ratio test.

5. Find the radius of convergence of

$$f(x) = \sum_{k=0}^{\infty} \frac{1}{2^k} x^{3k}$$

using the root test.

6. Find the radius of convergence of

$$f(x) = \sum_{k=0}^{\infty} \frac{1}{2^k} (x+1)^{3k}$$

using the ratio test.

7. Find the radius of convergence of

$$f(x) = \sum_{k=0}^{\infty} 2^k x^k$$

8. Use Sage to find the radius of convergence of

$$f(x) = \sum_{k=0}^{\infty} \frac{k^3+1}{3^k+1} x^{2k}$$

2.9.2 Part 2

In this part, we solve some differential equations using power series.

We want to solve a problem of the form

$$y''(x) + p(x)y'(x) + y(x) = f(x) \tag{2.25}$$

in the case where $p(x)$, $q(x)$ and $f(x)$ have a power series expansion. We will call a *power series solution* a series expansion for $y(x)$ where we have produced some algorithm or rule which enables us to compute as many of its coefficients as we like.

Solution strategy

Write $y(x) = a_0 + a_1 x + a_2 x^2 + \cdots = \sum_{k=0}^{\infty} a_k x^k$, for some real or complex numbers $a_0, a_1, \ldots$.

- Plug the power series expansions for y, p, q, and f into the differential equation (2.25).
- Comparing coefficients of like powers of x, derive relations between the a_j's.
- Using these recurrence relations [R-ps] and the initial conditions, solve for the coefficients of the power series of $y(x)$.

Example 2.9.4. Solve $y' - y = 5$, $y(0) = -4$, using the power series method.

This is easy to solve by undetermined coefficients: $y_h(x) = c_1 e^x$ and $y_p(x) = A_1$. Solving for A_1 gives $A_1 = -5$ and then solving for c_1 gives $-4 = y(0) = -5 + c_1 e^0$ so $c_1 = 1$ so $y = e^x - 5$.

Solving this using power series, we compute

$$\begin{array}{rcl} y'(x) & = & a_1 + 2a_2 x + 3a_3 x^2 + \cdots = \sum_{k=0}^{\infty} (k+1) a_{k+1} x^k, \\ -y(x) & = & -a_0 - a_1 x - a_2 x^2 - \cdots = \sum_{k=0}^{\infty} -a_k x^k, \\ ---- & -- & ------------------------------ \\ 5 & = & (-a_0 + a_1) + (-a_1 + 2a_2)x + \cdots = \sum_{k=0}^{\infty} (-a_k + (k+1)a_{k+1}) x^k. \end{array}$$

Comparing coefficients,

- for $k = 0$: $5 = -a_0 + a_1$;
- for $k = 1$: $0 = -a_1 + 2a_2$;
- for general k: $0 = -a_k + (k+1)a_{k+1}$ for $k > 0$.

The initial condition gives us $-4 = y(0) = a_0$, so

$$a_0 = -4, \quad a_1 = 1, \quad a_2 = 1/2, \quad a_3 = 1/6, \quad \cdots \quad , a_k = 1/k!.$$

This implies

$$y(x) = -4 + x + x/2 + \cdots + x^k/k! + \cdots = -5 + e^x,$$

which is in agreement with the previous discussion.

Example 2.9.5. Solve the differential equation

$$x^2y'' + xy' + x^2y = 0, \quad y(0) = 1, \quad y'(0) = 0,$$

using the power series method. This equation is called *Bessel's equation* [B-ps] of the zeroth order.

This differential equation is so well known (it has important applications to physics and engineering) that the coefficients in the series expansion were computed long ago. Most texts on special functions or differential equations have further details on this, but an online reference is [B-ps]. Its Taylor series expansion around 0 is

$$J_0(x) = \sum_{m=0}^{\infty} \frac{(-1)^m}{m!^2}\left(\frac{x}{2}\right)^{2m}$$

for all x. We shall next verify that $y(x) = J_0(x)$.

Let us try solving this ourselves using the power series method. We compute

$$\begin{array}{rcl}
x^2y''(x) & = & 0 + 0\cdot x + 2a_2x^2 + 6a_3x^3 + 12a_4x^4 + \cdots = \sum_{k=0}^{\infty}(k+2)(k+1)a_{k+2}x^k, \\
xy'(x) & = & 0 + a_1x + 2a_2x^2 + 3a_3x^3 + \cdots = \sum_{k=0}^{\infty} ka_kx^k, \\
x^2y(x) & = & 0 + 0\cdot x + a_0x^2 + a_1x^3 + \cdots = \sum_{k=2}^{\infty} a_{k-2}x^k, \\
---- & -- & ---------------------------- \\
0 & = & 0 + a_1x + (a_0 + 4a_2)x^2 + \cdots = a_1x + \sum_{k=2}^{\infty}(a_{k-2} + k^2a_k)x^k.
\end{array}$$

By the initial conditions, $a_0 = 1$, $a_1 = 0$. Comparing coefficients,

$$k^2a_k = -a_{k-2}, \quad k \geq 2,$$

which implies

$$a_2 = -\left(\frac{1}{2}\right)^2, \quad a_3 = 0, \quad a_4 = \left(\frac{1}{2}\cdot\frac{1}{4}\right)^2, \quad a_5 = 0, \quad a_6 = -\left(\frac{1}{2}\cdot\frac{1}{4}\cdot\frac{1}{6}\right)^2, \ldots.$$

In general,

$$a_{2k} = (-1)^k 2^{-2k}\frac{1}{k!^2}, \quad a_{2k+1} = 0,$$

for $k \geq 1$. This is in agreement with the series above for J_0.

Some of this computation can be formally done in `Sage` using power series rings.

Sage

```
sage: R6.<a0,a1,a2,a3,a4,a5,a6> = PolynomialRing(QQ,7)
sage: R.<x> = PowerSeriesRing(R6)
sage: y = a0 + a1*x + a2*x^2 + a3*x^3 + a4*x^4 + a5*x^5 +\
   a6*x^6 + O(x^7)
sage: y1 = y.derivative()
sage: y2 = y1.derivative()
sage: x^2*y2 + x*y1 + x^2*y
a1*x + (a0 + 4*a2)*x^2 + (a1 + 9*a3)*x^3 + (a2 + 16*a4)*x^4 +\
 (a3 + 25*a5)*x^5 + (a4 + 36*a6)*x^6 + O(x^7)
```

This is consistent with our computations above.

Sage knows quite a few special functions, such as the various types of Bessel functions.

Sage

```
sage: b = lambda x:bessel_J(x,0)
sage: P = plot(b,0,20,thickness=1)
sage: show(P)
sage: y = lambda x: 1 - (1/2)^2*x^2 + (1/8)^2*x^4 - (1/48)^2*x^6
sage: P1 = plot(y,0,4,thickness=1)
sage: P2 = plot(b,0,4,linestyle="--")
sage: show(P1+P2)
```

These are displayed in Figures 2.18 and 2.19.

Exercises

1. Using a power series around $t = 0$ of the form $y(t) = \sum_{n=0}^{\infty} c_n t^n$, find the recurrence relation for the c_i if y satisfies $y'' + ty' + y = 0$.

2. (a) Find the recurrence relation of the coefficients of the power series solution to the ODE $x' = 4x^3 x$ around $x = 0$, i.e., for solutions of the form $x = \sum_{n=0}^{\infty} c_n t^n$.

   (b) Find an explicit formula for c_n in terms of n.

3. Use Sage to find the first five terms in the power series solution to $x'' + x = 0$, $x(0) = 1$, $x'(0) = 0$. Plot this Taylor polynomial approximation over $-\pi < t < \pi$.

4. Find two linearly independent solutions of *Airy's equation* $x'' - tx = 0$ using power series.

5. Show that the coefficients of the power series solution to the initial value problem $y'' - y' - y = 0$, $y(0) = 0$, $y'(0) = 1$ have the form $c_n = F_n/n!$ where F_n is the nth Fibonacci number. (The Fibonacci numbers are $1, 1, 2, 3, 5, 8, 13, 21, 34, \ldots$, satisfying the recursion relation that each number is the sum of the previous two in the sequence.)

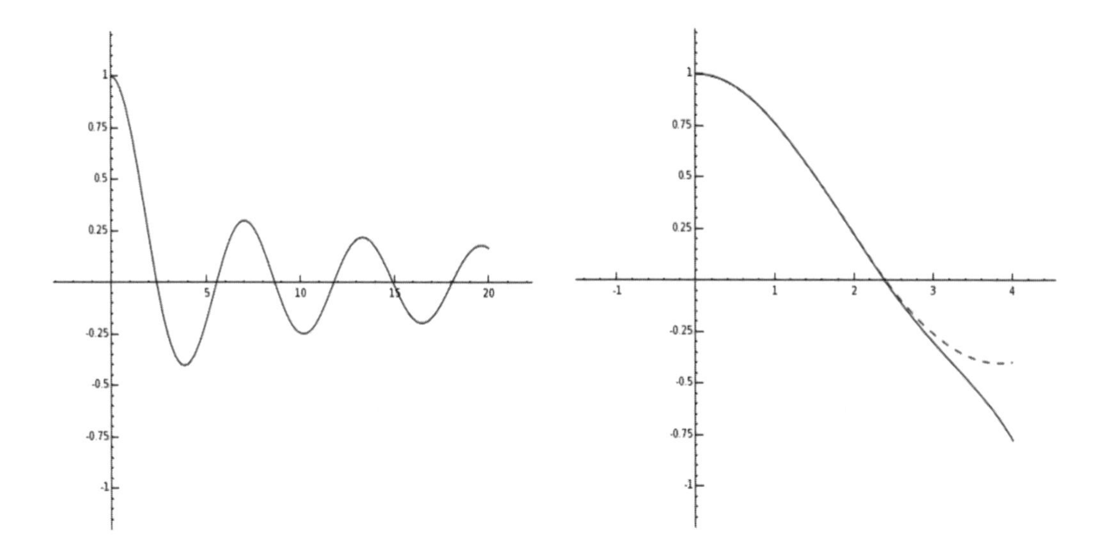

Figure 2.18: The Bessel function $J_0(x)$, for $0 < x < 20$.

Figure 2.19: A Taylor polynomial approximation for $J_0(x)$.

6. Determine the power series solution and radius of convergence of the ODE $y'' + x^2 y = 0$.

2.10 The Laplace transform method

What we know is not much. What we do not know is immense.

—*Pierre Simon Laplace*

Historically, the Laplace transform method of solving differential equations is attributed to Laplace himself. However, Oliver Heaviside[6] played an important role in his (independently made) rediscovery and implementation of the method.

2.10.1 Part 1

Pierre Simon Laplace (1749—1827) was a French mathematician and astronomer who is regarded as one of the greatest scientists of all time. His work was pivotal to the development of both celestial mechanics and probability [L-lt], [LT-lt].

The *Laplace transform* (sometimes abbreviated LT) of a function $f(t)$, defined for all real numbers $t \geq 0$, is the function $F(s)$, defined by

[6] A self-taught English electrical engineer, mathematician, and physicist, who lived from 1850 to 1925. Heaviside won the prestigious Faraday Medal in 1922.

$$F(s) = \mathcal{L}[f(t)] = \int_0^\infty e^{-st} f(t)\, dt.$$

The Laplace transform sends "nice" functions of t (we will be more precise later) to functions of another variable s. It has the wonderful property that it transforms constant-coefficient *differential* equations in t to *algebraic* equations in s.

The Laplace transform has two very familiar properties: Just as the integral of a sum is the sum of the integrals, the Laplace transform of a sum is the sum of Laplace transforms:

$$\mathcal{L}[f(t) + g(t)] = \mathcal{L}[f(t)] + \mathcal{L}[g(t)].$$

Just as constant factor can be taken outside of an integral, the Laplace transform of a constant times a function is that constant times the Laplace transform of that function:

$$\mathcal{L}[af(t)] = a\mathcal{L}[f(t)].$$

In other words, the Laplace transform is *linear.*

For which functions f is the Laplace transform actually defined? We want the indefinite integral above to converge (of course!). A function $f(t)$ is of *exponential order* α if there exist constants t_0 and M such that

$$|f(t)| < Me^{\alpha t} \quad \text{for all } t > t_0.$$

Roughly speaking, the magnitude of $f(t)$ should eventually be bounded above by some exponential function. If $\int_0^{t_0} f(t)\, dt$ exists and $f(t)$ is of exponential order α, then the Laplace transform $\mathcal{L}[f](s)$ exists for $s > \alpha$.

Example 2.10.1. Consider the Laplace transform of $f(t) = 1$. The Laplace transform integral converges for $s > 0$:

$$\begin{aligned}\mathcal{L}[f](s) &= \int_0^\infty e^{-st}\, dt \\ &= \left[-\frac{1}{s}e^{-st}\right]_0^\infty \\ &= \frac{1}{s}.\end{aligned}$$

Example 2.10.2. Consider the Laplace transform of $f(t) = e^{at}$. The Laplace transform integral converges for $s > a$:

$$\begin{aligned}\mathcal{L}[f](s) &= \int_0^\infty e^{(a-s)t}\, dt \\ &= \left[-\frac{1}{s-a}e^{(a-s)t}\right]_0^\infty \\ &= \frac{1}{s-a}.\end{aligned}$$

Example 2.10.3. Consider the Laplace transform of the translated *unit step* (or Heaviside) function,

$$u(t-c) = \begin{cases} 0 & \text{for } t < c \\ 1 & \text{for } t > c, \end{cases}$$

where $c > 0$ (this function is sometimes also denoted $H(t-c)$). This function is "off" (i.e., equal to 0) until you get to $t = c$, at which time it turns "on." Its Laplace transform is

$$\begin{aligned} \mathcal{L}[u(t-c)] &= \int_0^\infty e^{-st} u(t-c)\, dt \\ &= \int_c^\infty e^{-st}\, dt \\ &= \left[\frac{e^{-st}}{-s}\right]_c^\infty \\ &= \frac{e^{-cs}}{s} \quad \text{for } s > 0. \end{aligned}$$

Using this formula, any linear combination of translates of the unit step function can also be computed. For example, if

$$f(t) = 2u(t) - 3u(t-2) + 5u(t-6),$$

then

$$F(s) = \frac{2}{s} - \frac{3e^{-2s}}{s} + \frac{5e^{-6s}}{s}.$$

The *inverse Laplace transform* is denoted

$$f(t) = \mathcal{L}^{-1}[F(s)](t),$$

where $F(s) = \mathcal{L}[f(t)](s)$.

Usually we compute the inverse Laplace transform by looking up the form of the transform in a table. It is possible to compute the inverse directly with complex-valued integrals, which is a topic in a more advanced course in complex analysis. There is also the amazing (although impractical) formula of Emil Post [Post]:

$$f(t) = \mathcal{L}^{-1}[F(s)] = \lim_{n\to\infty} \frac{(-1)^n}{n!} \left(\frac{n}{t}\right)^{n+1} F^{(n)}\left(\frac{n}{t}\right). \tag{2.26}$$

Example 2.10.4. Consider

$$f(t) = \begin{cases} 1 & \text{for } t < 2, \\ 0 & \text{on } t \geq 2. \end{cases}$$

(Incidentally, this can also be written $1 - u(t-2)$.) We show how Sage can be used to compute the Laplace transform of this.

Sage

```
sage: t = var('t')
sage: s = var('s')
sage: f = Piecewise([[(0,2),1],[(2,infinity),0]])
sage: f.laplace(t, s)
1/s - e^(-(2*s))/s
sage: f1 = lambda t: 1
sage: f2 = lambda t: 0
sage: f = Piecewise([[(0,2),f1],[(2,10),f2]])
sage: P = f.plot(rgbcolor=(0.7,0.1,0.5),thickness=3)
sage: show(P)
```

According to Sage, $\mathcal{L}[f](s) = 1/s - e^{-2s}/s$. Note that the function f was redefined for plotting purposes only (the fact that it was redefined over $0 < t < 10$ means that Sage will plot it over that range.) The plot of this function is displayed in Figure 2.20.

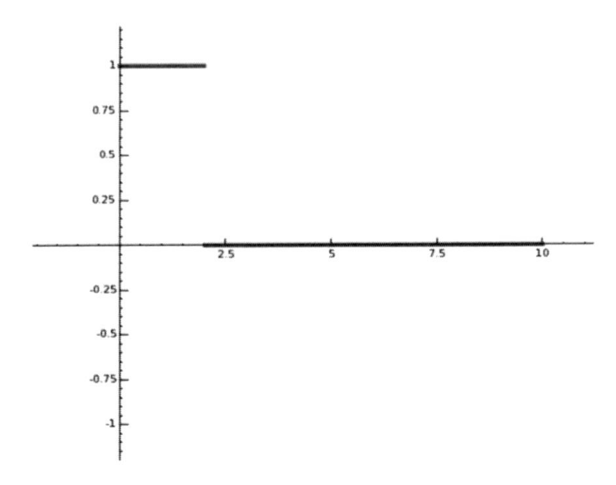

Figure 2.20: The piecewise constant function $1 - u(t-2)$.

Next, some properties of the Laplace transform.

- Differentiate the definition of the Laplace transform with respect to s:

$$F'(s) = -\int_0^\infty e^{-st} t f(t)\, dt.$$

Repeating this over and over again gives

$$\frac{d^n}{ds^n} F(s) = (-1)^n \int_0^\infty e^{-st} t^n f(t)\, dt. \tag{2.27}$$

- In the definition of the Laplace transform, replace $f(t)$ by its derivative $f'(t)$:

$$\mathcal{L}\left[f'(t)\right](s) = \int_0^\infty e^{-st} f'(t)\, dt.$$

Now integrate by parts ($u = e^{-st}$, $dv = f'(t)\, dt$), to obtain

$$\int_0^\infty e^{-st} f'(t)\, dt = f(t)e^{-st}|_0^\infty - \int_0^\infty f(t) \cdot (-s) \cdot e^{-st}\, dt = -f(0) + s\mathcal{L}\left[f(t)\right](s).$$

In other words, if $F(s)$ is the Laplace transform of $f(t)$ then $sF(s) - f(0)$ is the Laplace transform of $f'(t)$:

$$\mathcal{L}\left[f'(t)\right](s) = s\mathcal{L}\left[f(t)\right](s) - f(0). \tag{2.28}$$

- If you replace f by f' in (2.28),

$$\mathcal{L}\left[f''(t)\right](s) = s\mathcal{L}\left[f'(t)\right](s) - f'(0), \tag{2.29}$$

and apply (2.28) again, then you will get

$$\mathcal{L}\left[f''(t)\right](s) = s^2\mathcal{L}\left[f(t)\right](s) - sf(0) - f'(0). \tag{2.30}$$

- Using (2.28) and (2.30), the Laplace transform of any constant-coefficient ODE

$$ax''(t) + bx'(t) + cx(t) = f(t)$$

is

$$a(s^2\mathcal{L}\left[x(t)\right](s) - sx(0) - x'(0)) + b(s\mathcal{L}\left[x(t)\right](s) - x(0)) + c\mathcal{L}\left[x(t)\right](s) = F(s),$$

where $F(s) = \mathcal{L}\left[f(t)\right](s)$. In particular, the Laplace transform of the solution, $X(s) = \mathcal{L}\left[x(t)\right](s)$, satisfies

$$X(s) = (F(s) + ax(0)s + ax'(0) + bx(0))/(as^2 + bs + c).$$

Note that the denominator is the characteristic polynomial of the differential equation.

Moral of the story: Using Laplace transforms, you can replace any constant coefficient nonhomogeneous linear differential equation for $x = x(t)$ by an algebraic equation for $X = X(s)$. It is generally very easy to compute the Laplace transform $X(s)$ of the solution to any constant coefficient nonhomogeneous linear ODE. Computing the actual solution $x(t)$ is usually much more work.

Example 2.10.5. We know now how to compute not only the Laplace transform of $f(t) = e^{at}$ (it's $F(s) = (s-a)^{-1}$) but also the Laplace transform of any function of the form $t^n e^{at}$ by differentiating it:

$$\mathcal{L}\left[te^{at}\right] = -F'(s) = (s-a)^{-2}, \quad \mathcal{L}\left[t^2e^{at}\right] = F''(s) = 2 \cdot (s-a)^{-3}, \quad \dots,$$

and in general

$$\mathcal{L}\left[t^n e^{at}\right] = -F'(s) = n! \cdot (s-a)^{-n-1}. \tag{2.31}$$

Let us now solve a first-order ODE using Laplace transforms.

Example 2.10.6. Let us solve the differential equation

$$x' + x = t^{100}e^{-t}, \quad x(0) = 0$$

using Laplace transforms. Note that it would be highly impractical to solve this using undetermined coefficients. (You would have 101 undetermined coefficients to solve for!)

First, we compute the Laplace transform of the solution to the differential equation. The Laplace transform of the left-hand side, by (2.31), is

$$\mathcal{L}\left[x' + x\right] = sX(s) - x(0) + X(s) = (s+1)X(s),$$

where $F(s) = \mathcal{L}\left[f(t)\right](s)$. For the Laplace transform of the right-hand side, let

$$F(s) = \mathcal{L}\left[e^{-t}\right] = \frac{1}{s+1}.$$

By (2.27),

$$\frac{d^{100}}{ds^{100}}F(s) = \mathcal{L}\left[t^{100}e^{-t}\right] = \frac{d^{100}}{ds^{100}}\frac{1}{s+1}.$$

The first several derivatives of $\frac{1}{s+1}$ are as follows:

$$\frac{d}{ds}\frac{1}{s+1} = -\frac{1}{(s+1)^2}, \quad \frac{d^2}{ds^2}\frac{1}{s+1} = 2\frac{1}{(s+1)^3}, \quad \frac{d^3}{ds^3}\frac{1}{s+1} = -62\frac{1}{(s+1)^4},$$

and so on. Therefore, the Laplace transform of the right-hand side is:

$$\frac{d^{100}}{ds^{100}}\frac{1}{s+1} = 100!\frac{1}{(s+1)^{101}}.$$

Consequently,

$$X(s) = 100!\frac{1}{(s+1)^{102}}.$$

Using (2.31), we can compute the inverse Laplace transform of this:

$$x(t) = \mathcal{L}^{-1}\left[X(s)\right] = \mathcal{L}^{-1}\left[100!\frac{1}{(s+1)^{102}}\right] = \frac{1}{101}\mathcal{L}^{-1}\left[101!\frac{1}{(s+1)^{102}}\right] = \frac{1}{101}t^{101}e^{-t}.$$

Let us now solve a second-order ODE using Laplace transforms.

Example 2.10.7. Let us solve the differential equation

$$x'' + 2x' + 2x = e^{-2t}, \quad x(0) = x'(0) = 0$$

using Laplace transforms.

The Laplace transform of the left-hand side, by (2.31) and (2.29), is

$$\mathcal{L}\left[x'' + 2x' + 2x\right] = (s^2 + 2s + 2)X(s),$$

as in the previous example. The Laplace transform of the right-hand side is

$$\mathcal{L}\left[e^{-2t}\right] = \frac{1}{s+2}.$$

Solving for the Laplace transform of the solution algebraically:

$$X(s) = \frac{1}{(s+2)((s+1)^2+1)}.$$

The inverse Laplace transform of this can be obtained from Laplace transform tables after rewriting it using partial fractions:

$$X(s) = \frac{1}{2}\cdot\frac{1}{s+2} - \frac{1}{2}\frac{s}{(s+1)^2+1} = \frac{1}{2}\cdot\frac{1}{s+2} - \frac{1}{2}\frac{s+1}{(s+1)^2+1} + \frac{1}{2}\frac{1}{(s+1)^2+1}.$$

The inverse Laplace transform is:

$$x(t) = \mathcal{L}^{-1}\left[X(s)\right] = \frac{1}{2}\cdot e^{-2t} - \frac{1}{2}\cdot e^{-t}\cos(t) + \frac{1}{2}\cdot e^{-t}\sin(t).$$

Next, we show how `Sage` can be used to do some of this. We break the `Sage` solution into steps.

Step 1. First, we type in the ODE and take its Laplace transform.

Sage

```
sage: s,t,X = var('s,t,X')
sage: x = function("x",t)
sage: de = diff(x,t,t)+2*diff(x,t)+2*x==e^(-2*t)
sage: laplace(de,t,s)
s^2*laplace(x(t), t, s) + 2*s*laplace(x(t), t, s) - s*x(0)
   + 2*laplace(x(t), t, s) - 2*x(0) - D[0](x)(0) == (1/(s + 2))
sage: LTde = laplace(de,t,s)
```

Step 2. Now we solve this equation for `X = laplace(x(t), t, s)`. For this, we use Python to do some string replacements. Python is the underlying language for `Sage` and has very powerful string manipulation functions.

Sage

```
sage: strLTde = str(LTde).replace("laplace(x(t), t, s)",  "X")
sage: strLTde0 = strLTde.replace("x(0)","0")
sage: strLTde00 = strLTde0.replace("D[0](x)(0)","0")
sage: LTde00 = sage_eval(strLTde00,locals={"s":s,"X":X})
sage: soln = solve(LTde00,X)
sage: Xs = soln[0].rhs(); Xs
1/(s^3 + 4*s^2 + 6*s + 4)
```

Step 3. Now that we have solved for the Laplace transform of the solution, we take inverse Laplace transforms to get the solution to the original ODE. There are various ways to do this. One way that is convenient if you also want to check the answer using tables, is to compute the partial fraction decomposition and then take inverse Laplace transforms. Of course, you can also use **Sage** to compute the inverse Laplace transform.

Sage

```
sage: factor(s^3 + 4*s^2 + 6*s + 4)
(s + 2)*(s^2 + 2*s + 2)
sage: f = 1/((s+2)*((s+1)^2+1))
sage: f.partial_fraction()
 1/(2*(s + 2)) - s/(2*(s^2 + 2*s + 2))
sage: f.inverse_laplace(s,t)
 e^(-t)*(sin(t)/2 - cos(t)/2) + e^(-(2*t))/2
```

Exercises

1. Compute the Laplace transform of the function

$$v(t) = \begin{cases} 1 & \text{for } t \in [0,1], \\ 0 & \text{for } t \notin [0,1] \end{cases}$$

   directly from the definition $\mathcal{L}(v) = \int_0^\infty e^{-st} v(t)dt$.

2. Write

$$f(t) = \begin{cases} -1 & \text{for } t \in [0,2], \\ 3 & \text{for } t \in (2,5], \\ 2 & \text{for } t \in (5,7], \\ 0 & \text{for } t \notin [0,7] \end{cases}$$

   as a linear combination of translates of the unit step function. Compute the Laplace transform of the function using the formula in Example 2.10.3.

3. Compute the Laplace transform of $f(t) = 1 + \cos(3t)$.

4. Compute the Laplace transform of $f(t) = \sin(2t + \pi/4)$.

5. Compute the Laplace transform of $f(t) = (t+1)^2$.

6. Compute the Laplace transform of $f(t) = t - 3e^{3t}$.

7. Compute the Laplace transform of $f(t) = 2te^t$.

8. Compute the Laplace transform of $f(t) = te^t \sin(t)$.

9. Use Sage to solve the differential equation below using the Laplace transform method.

$$x'' + 2x' + 5x = e^{-t}, \quad x(0) = x'(0) = 0,$$

as in the example above.

10. Use the Laplace transform method to solve the initial value problem $x'' - x' - 2x = 0$, $x(0) = 0$, $x'(0) = 1$.

2.10.2 Part 2

In this part, we shall focus on two other aspects of Laplace transforms:

- solutions of differential equations involving translations of the unit step (Heaviside) function,
- convolutions and applications.

It follows from the definition of the Laplace transform that if

$$f(t) \stackrel{\mathcal{L}}{\longmapsto} F(s) = \mathcal{L}[f(t)](s),$$

then

$$f(t)u(t-c) \stackrel{\mathcal{L}}{\longmapsto} e^{-cs}\mathcal{L}[f(t+c)](s) \tag{2.32}$$

and

$$f(t-c)u(t-c) \stackrel{\mathcal{L}}{\longmapsto} e^{-cs}F(s). \tag{2.33}$$

These two properties are called *translation theorems.*

Example 2.10.8. First, consider the Laplace transform of the piecewise-defined function $f(t) = (t-1)^2 u(t-1)$. Using (2.33), this is

$$\mathcal{L}[f(t)] = e^{-s}\mathcal{L}[t^2](s) = 2\frac{1}{s^3}e^{-s}.$$

Second, consider the Laplace transform of the piecewise-constant function

$$f(t) = \begin{cases} 0 & \text{for } t < 0, \\ -1 & \text{for } 0 \le t \le 2, \\ 1 & \text{for } t > 2. \end{cases}$$

This can be expressed as $f(t) = -u(t) + 2u(t-2)$, so

$$\begin{aligned} \mathcal{L}[f(t)] &= -\mathcal{L}[u(t)] + 2\mathcal{L}[u(t-2)] \\ &= -\frac{1}{s} + 2\frac{1}{s}e^{-2s}. \end{aligned}$$

Finally, consider the Laplace transform of $f(t) = \sin(t)u(t-\pi)$. Using (2.32), this is

$$\mathcal{L}[f(t)] = e^{-\pi s}\mathcal{L}[\sin(t+\pi)](s) = e^{-\pi s}\mathcal{L}[-\sin(t)](s) = -e^{-\pi s}\frac{1}{s^2+1}.$$

The plot of this function $f(t) = \sin(t)u(t-\pi)$ is displayed in Figure 2.21.

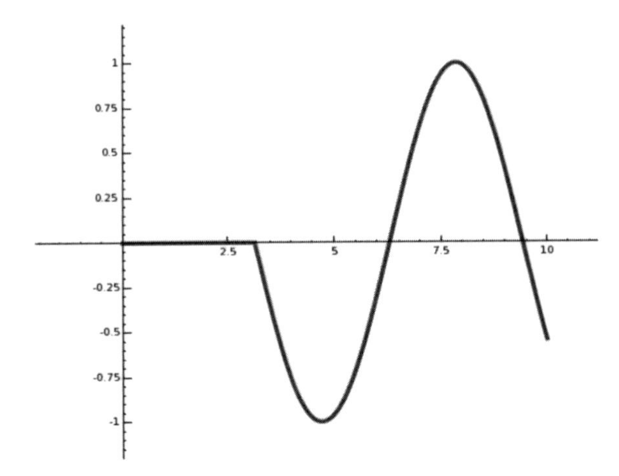

Figure 2.21: The piecewise-continuous function $u(t-\pi)\sin(t)$.

We show how **Sage** can be used to compute these Laplace transforms.

Sage

```
sage: t = var('t')
sage: s = var('s')
sage: assume(s>0)
sage: f = Piecewise([[(0,1),0],[(1,infinity),(t-1)^2]])
sage: f.laplace(t, s)
2*e^(-s)/s^3
sage: f = Piecewise([[(0,2),-1],[(2,infinity),2]])
sage: f.laplace(t, s)
```

```
3*e^(-(2*s))/s - 1/s
sage: f = Piecewise([[(0,pi),0],[(pi,infinity),sin(t)]])
sage: f.laplace(t, s)
-e^(-(pi*s))/(s^2 + 1)
sage: f1 = lambda t: 0
sage: f2 = lambda t: sin(t)
sage: f = Piecewise([[(0,pi),f1],[(pi,10),f2]])
sage: P = f.plot(rgbcolor=(0.7,0.1,0.5),thickness=3)
sage: show(P)
```

The plot given by these last few commands is displayed in Figure 2.21.

Before turning to differential equations, let us introduce convolutions.

Let $f(t)$ and $g(t)$ be continuous (for $t \geq 0$; for $t < 0$, we assume $f(t) = g(t) = 0$). The *convolution* of $f(t)$ and $g(t)$ is defined by

$$(f * g) = \int_0^t f(z)g(t-z)\,dz = \int_0^t f(t-z)g(z)\,dz.$$

The *convolution theorem* states

$$\mathcal{L}[f * g(t)](s) = F(s)G(s) = \mathcal{L}[f](s)\mathcal{L}[g](s).$$

The Laplace transform of the convolution is the product of the Laplace transforms. (Or, equivalently, the inverse Laplace transform of the product is the convolution of the inverse Laplace transforms.)

To show this, do a change of variables in the following double integral:

$$\begin{aligned}
\mathcal{L}[f * g(t)](s) &= \int_0^\infty e^{-st} \int_0^t f(z)g(t-z)\,dz\,dt \\
&= \int_0^\infty \int_z^\infty e^{-st} f(z)g(t-z)\,dt\,dz \\
&= \int_0^\infty e^{-sz} f(z) \int_z^\infty e^{-s(t-z)} g(t-z)\,dt\,dz \\
&= \int_0^\infty e^{-su} f(u)\,du \int_0^\infty e^{-sv} g(v)\,dv \\
&= \mathcal{L}[f](s)\mathcal{L}[g](s).
\end{aligned}$$

Example 2.10.9. Consider the inverse Laplace transform of $\frac{1}{s^3-s^2}$. This can be computed using partial fractions and Laplace transform tables.[7] However, it can also be computed using convolutions.

First we factor the denominator, as follows

[7] There are Laplace transforms in Table 2.2.

$$\frac{1}{s^3 - s^2} = \frac{1}{s^2}\frac{1}{s-1}.$$

We know the inverse Laplace transforms of each term:

$$\mathcal{L}^{-1}\left[\frac{1}{s^2}\right] = t, \qquad \mathcal{L}^{-1}\left[\frac{1}{s-1}\right] = e^t.$$

We apply the convolution theorem:

$$\begin{aligned}\mathcal{L}^{-1}\left[\frac{1}{s^2}\frac{1}{s-1}\right] &= \int_0^t ze^{t-z}\,dz \\ &= e^t\left[-ze^{-z}\right]_0^t - e^t\int_0^t -e^{-z}\,dz \\ &= -t - 1 + e^t.\end{aligned}$$

Therefore,

$$\mathcal{L}^{-1}\left[\frac{1}{s^2}\frac{1}{s-1}\right](t) = e^t - t - 1.$$

Example 2.10.10. Here is a neat application of the convolution theorem. Consider the convolution

$$f(t) = 1 * 1 * 1 * 1 * 1.$$

What is it? First, take the Laplace transform. Since the Laplace transform of the convolution is the product of the Laplace transforms:

$$\mathcal{L}[1 * 1 * 1 * 1 * 1](s) = (1/s)^5 = \frac{1}{s^5} = F(s).$$

We know from Laplace transform tables that $\mathcal{L}^{-1}\left[\frac{4!}{s^5}\right](t) = t^4$, so

$$f(t) = \mathcal{L}^{-1}\left[F(s)\right](t) = \frac{1}{4!}\mathcal{L}^{-1}\left[\frac{4!}{s^5}\right](t) = \frac{1}{4!}t^4.$$

This clever computation is called "computing $1*1*1*1*1$ using the convolution theorem."

You can also compute $1 * 1 * 1 * 1 * 1$ directly in Sage:

Sage

```
sage: t,z = var("t,z")
sage: f(t) = 1
sage: ff = integral(f(t-z)*f(z),z,0,t); ff
t
sage: fff = integral(f(t-z)*ff(z),z,0,t); fff
1/2*t^2
sage: ffff = integral(f(t-z)*fff(z),z,0,t); ffff
```

```
1/6*t^3
sage: fffff = integral(f(t-z)*ffff(z),z,0,t); fffff
1/24*t^4
sage: s = var("s")
sage: (1/s^5).inverse_laplace(s,t)
1/24*t^4
```

Example 2.10.11. Compute $\sin(t) * \cos(t)$ using the convolution theorem.

Let $f(t) = \sin(t) * \cos(t)$. By the convolution theorem, we have

$$F(s) = \frac{s}{(s^2+1)^2}.$$

The inverse Laplace transform can be computed using Laplace transform tables:

$$\mathcal{L}^{-1}[F(s)](t) = \frac{1}{2} t \sin(t).$$

Example 2.10.12. Here is another application of the convolution theorem. Consider the convolution

$$f(t) = \sin(t) * u(t-1).$$

What is it? First, take the Laplace transform, using the convolution theorem:

$$F(s) = \frac{e^{-s}}{s^3+s}.$$

By the translation theorem (2.33), the inverse Laplace transform of $F(s)$ is the translation of the inverse Laplace transform of $(s^3+s)^{-1} = -s/(s^2+1) + 1/s$:

$$f(t) = (-\cos(t-1)+1)u(t-1).$$

A general method

Now let us turn to solving a second-order constant-coefficient differential equation,

$$ay'' + by' + cy = f(t), \quad y(0) = y_0, \quad y'(0) = y_1. \tag{2.34}$$

First, take Laplace transforms of both sides:

$$as^2Y(s) - asy_0 - ay_1 + bsY(s) - by_0 + cY(s) = F(s),$$

so

$$Y(s) = \frac{1}{as^2+bs+c}F(s) + \frac{asy_0 + ay_1 + by_0}{as^2+bs+c}. \tag{2.35}$$

The function $\frac{1}{as^2+bs+c}$ is sometimes called the *transfer function* (this is an engineering term) and its inverse Laplace transform,

$$w(t) = \mathcal{L}^{-1}\left[\frac{1}{as^2+bs+c}\right](t),$$

the *weight function* for the differential equation. Roughly speaking, this represents the "response" to a "unit impulse" at $t = 0$.

Lemma 2.10.1. *If $a \neq 0$ then $w(t) = 0$.*

(The only proof we have of this is a case-by-case proof using Laplace transform tables. Case 1 is when the roots of $as^2 + bs + c = 0$ are real and distinct, case 2 is when the roots are real and repeated, and case 3 is when the roots are complex. In each case, $w(0) = 0$. The verification of this is left to the reader, if he or she is interested.)

By the above lemma and the first derivative theorem,

$$w'(t) = \mathcal{L}^{-1}\left[\frac{s}{as^2+bs+c}\right](t).$$

Using this and the convolution theorem, the inverse Laplace transform of (2.35) is

$$y(t) = (w * f)(t) + ay_0 \cdot w'(t) + (ay_1 + by_0) \cdot w(t). \tag{2.36}$$

This proves the following fact, sometimes referred to as the *impulse response formula.*

Theorem 2.10.1. *The unique solution to the differential equation (2.34) is (2.36).*

Example 2.10.13. Consider the differential equation $y'' + y = 1$, $y(0) = y'(0) = 0$.

The weight function is the inverse Laplace transform of $\frac{1}{s^2+1}$, so $w(t) = \sin(t)$. By (2.36),

$$y(t) = 1 * \sin(t) = \int_0^t \sin(u)\, du = -\cos(u)|_0^t = 1 - \cos(t).$$

(Yes, it is just that easy to solve that diferential equation.)

If the "impulse" $f(t)$ is piecewise-defined, sometimes the convolution term in the formula (2.36) is awkward to compute.

Example 2.10.14. Consider the differential equation $y'' + y = u(t-1)$, $y(0) = y'(0) = 0$.

The weight function is the inverse Laplace transform of $\frac{1}{s^2+1}$, so $w(t) = \sin(t)$. By (2.36),

$$y(t) = \sin(t) * u(t-1) = (-\cos(t-1) + 1)u(t-1),$$

by Example 2.10.12.

Example 2.10.15. Consider the differential equation $y'' - y' = u(t-1)$, $y(0) = y'(0) = 0$.

Taking Laplace transforms gives $s^2Y(s) - sY(s) = \frac{1}{s}e^{-s}$, so

$$Y(s) = \frac{1}{s^3 - s^2}e^{-s}.$$

We know from a previous example that

$$\mathcal{L}^{-1}\left[\frac{1}{s^3 - s^2}\right](t) = e^t - t - 1,$$

so by the translation theorem (2.33), we have

$$y(t) = \mathcal{L}^{-1}\left[\frac{1}{s^3 - s^2}e^{-s}\right](t) = (e^{t-1} - (t-1) - 1) \cdot u(t-1) = (e^{t-1} - t) \cdot u(t-1).$$

Example 2.10.16. We try to solve $x'' - x = u(t-1)$, $x(0) = x'(0) = 0$ using Sage.

At this stage, Sage lacks the functionality to solve this differential equation as easily as others but we can still use Sage to help with the solution.

Sage

```
sage: t = var('t')
sage: x = function('x', t)
sage: u1 = lambda t: (sign(t-1)+1)/2
sage: de = lambda y: diff(y,t,t) - y - u2(t)
sage: desolve(de(x),[x,t])
1/4*(e^(2*t)*integrate(e^(-t)*sgn(t - 2), t) - 2*e^t -
  integrate(e^t*sgn(t - 2), t))*e^(-t) + k1*e^t + k2*e^(-t)
```

You see that some terms are unevaluated integrals.

We try another approach. First, we initialize some variables and take the Laplace transform of $f(t) = u(t-1)$.

Sage

```
sage: s,t,X = var('s,t,X')
sage: x = function("x",t)
sage: f1 = 0
sage: f2 = 1
sage: u1 = Piecewise([[(0,1),f1],[(1,Infinity),f2]])
sage: F = u1.laplace(t,s); F
e^(-s)/s
```

Next, we take the Laplace transform of the differential equation

$$x'' - x = f(t),$$

with an arbitrary function $f(t)$.

Sage

```
sage: ft = function("ft",t)
sage: de = diff(x,t,t) - x==ft
```

```
sage: LTde = laplace(de,t,s); LTde
s^2*laplace(x(t), t, s) - s*x(0) - laplace(x(t), t, s) - D[0](x)(0) ==
laplace(ft(t), t, s)
```

Next, we take this equation and solve it for $X(s)$ using Python's string manipulation functionality:

Sage

```
sage: strLTde = str(LTde).replace("laplace(x(t), t, s)",  "X")
sage: strLTde0 = strLTde.replace("x(0)","0")
sage: strLTde00 = strLTde0.replace("D[0](x)(0)","0")
sage: strLTde00F = strLTde00.replace("laplace(ft(t), t, s)",  "F")
sage: strLTde00F
's^2*X - s*0 - X - 0 == F'
```

Next,we plug in for $f(t)$ the function $u(t-1)$ (denoted `u1` in the above **Sage** code) and solve:

Sage

```
sage: LTde00F = sage_eval(strLTde00F,locals={"s":s,"X":X,"F":F})
sage: LTde00F
X*s^2 - X == e^(-s)/s
sage: soln = solve(LTde00F,X)
sage: Xs = soln[0].rhs(); Xs
e^(-s)/(s^3 - s)
sage: Xs.partial_fraction()
1/2*e^(-s)/(s - 1) + 1/2*e^(-s)/(s + 1) - e^(-s)/s
```

Unfortunately, at this time, **Sage** cannot take the inverse Laplace transform of this. We can look up the terms in Laplace transform tables (such as Table 2.2) and do this by hand with ease:

$$x(t) = \frac{1}{2}e^{t-1}u(t-1) + \frac{1}{2}e^{-(t-1)}u(t-1) - u(t-1).$$

This solves $x'' - x = u(t-1)$, $x(0) = x'(0) = 0$.

Dirac delta functions

The "Dirac delta function" $\delta(t)$ is technically not a function. Roughly speaking, it may be thought of as being analogous to a radar "ping": if a continuous function $f(t)$ represents an objects' trajectory then the delta function "samples" its value at $t = 0$. The "graph" of $\delta(t)$ can be visualized as in Figure 2.22.

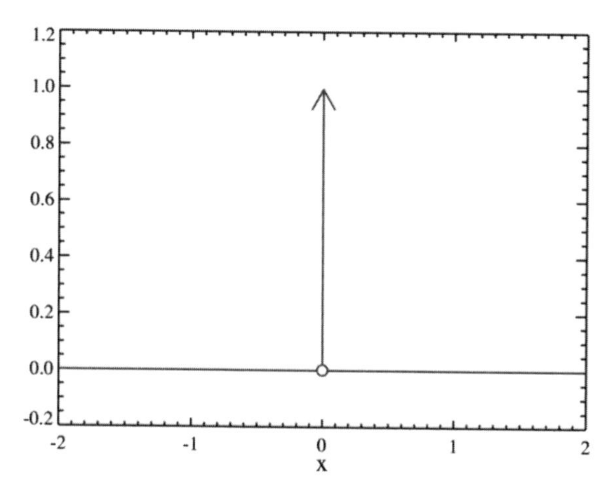

Figure 2.22: The delta function.

To be precise, $\delta(t)$ is a linear functional (that is, a linear function which sends functions to numbers) defined by the formula

$$\int_{\infty}^{\infty} \delta(t) f(t)\, dt = f(0),$$

where $f(t)$ is any continuous function. Of course, the interval $(-\infty, \infty)$ can be replaced by any interval containing 0. In particular, if $a > 0$ then

$$\int_0^{\infty} \delta(t-a) f(t) e^{-st}\, dt = f(a) e^{-as}.$$

A special case of the above formula gives the Laplace transform of the delta function:

$$\mathcal{L}(\delta(t)) = \int_0^{\infty} \delta(t) e^{-st}\, dt = 1.$$

Though it is not obvious at first glance, the delta function is not unrelated to the unit step function. Indeed, in some sense the delta function may be regarded as the derivative of the (discontinuous!) unit step function. Let $f(t)$ be any function on the real line whose derivative is integrable on $(0, \infty)$ and satisfies $\lim_{t\to\infty} f(t) = 0$. Formally integrating by parts, we obtain

$$\begin{aligned} -\textstyle\int_{\infty}^{\infty} \delta(t) f(t)\, dt &= -f(0) = \textstyle\int_0^{\infty} f'(t)\, dt = \int_{-\infty}^{\infty} u(t) f'(t)\, dt \\ &= u(t) f(t)|_{-\infty}^{\infty} - \textstyle\int_{-\infty}^{\infty} u'(t) f(t)\, dt = -\int_{-\infty}^{\infty} u'(t) f(t)\, dt. \end{aligned}$$

Comparing the left-hand side with the right-hand side, we see that in some sense (which can be made precise using the theory of functional analysis, but which goes beyond the scope of this book), we have

$$\delta(t) = u'(t).$$

Example 2.10.17. Consider the differential equation $x''+x=-\delta(t-\pi)$, $x(0)=x'(0)=0$. This models a unit mass attached to an undamped spring suspended from a board with no initial displacement or initial velocity. At time $t\pi$, the board is hit very hard (say with a hammer blow) in the upward direction. As we will see, this will start the mass oscillating.

Taking Laplace transforms gives $s^2X(s)+X(s)=e^{-\pi s}$, so

$$X(s) = -\frac{1}{s^2+1}e^{-\pi s}.$$

The inverse Laplace transform is $x(t)=\sin(t)u(t-\pi)$ (see Example 2.10.8, which also has a graph of this solution).

Function	Transform
1	$\frac{1}{s}$
t	$\frac{1}{s^2}$
t^n	$\frac{n!}{s^{n+1}}$
e^{at}	$\frac{1}{s-a}$
$\cos(kt)$	$\frac{s}{s^2+k^2}$
$\sin(kt)$	$\frac{k}{s^2+k^2}$
$e^{at}\sin(kt)$	$\frac{k}{(s-a)^2+k^2}$
$e^{at}\cos(kt)$	$\frac{s}{(s-a)^2+k^2}$
$e^{at}f(t)$	$F(s-a)$
$-tf(t)$	$F'(s)$
$\int_0^t f(\tau)d\tau$	$F(s)/s$
$f'(t)$	$sF(s)-f(0)$
$f''(t)$	$s^2F(s)-sf(0)-f'(0)$
$f*g(t)$	$F(s)G(s)$
$\delta(t)$	1

Table 2.2: Some Laplace transforms, $\mathcal{L}[f(t)](s)=F(s)$.

2.10.3 Part 3

We conclude our study of the Laplace transform with the case of a periodic nonhomogeneous term, i.e., ODEs of the form

$$ax''(t)+bx'(t)+cx(t)=f(t)$$

where $f(t)$ is a periodic function of period T. This means that $f(t)=f(t+T)$ for all t. For this important special case we can reduce the improper integral in the Laplace transform to the finite interval $[0,T]$.

!h

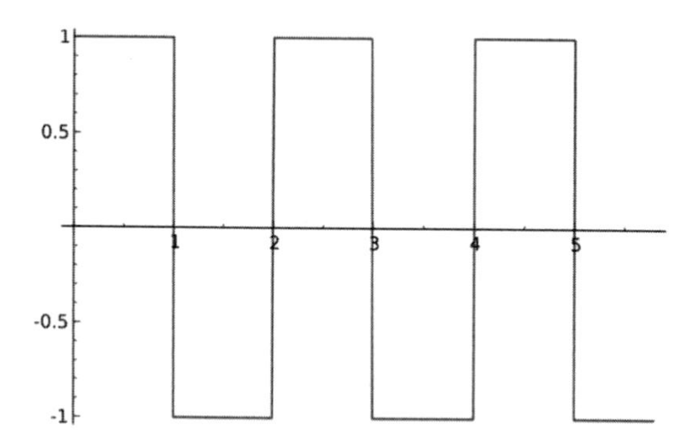

Figure 2.23: The square-wave function

Lemma 2.10.2. *If $f(t)$ is a periodic function with period T, then*

$$\mathcal{L}[f(t)](s) = \frac{1}{1-e^{-Ts}} \int_0^T e^{-st} f(t)dt.$$

Proof. First we write the Laplace transform as a sum over the period intervals:

$$\mathcal{L}[f(t)](s) = \int_0^\infty e^{-st} f(t)dt = \sum_{n=0}^\infty \int_{nT}^{(n+1)T} e^{-st} f(t)dt,$$

and then use the substitution $u = t - nT$ in each interval. For convenience, we then relabel u as t once again, to get

$$\mathcal{L}[f(t)](s) = \sum_{n=0}^\infty \int_0^T e^{-s(t-nT)} f(t)\ dt = \left(\int_0^T e^{-st} f(t)\ dt\right) \sum_{n=0}^\infty e^{-nsT}.$$

The sum is over a geometric series in e^{-sT} since $e^{-nsT} = (e^{-sT})^n$. For $s > 0$, $e^{-sT} < 1$ and the series converges ($\sum_{n=0}^\infty e^{-nsT} = \frac{1}{1-e^{-Ts}}$), giving the desired result. □

Exercises

1. Compute the convolution $t * t$.

2. Compute the convolution $e^t * e^t$.

3. Compute the convolution $\sin(t) * \sin(t)$.

4. Show that the Laplace transform of the square-wave function $g(t)$ graphed in Figure 2.23 is equal to $\dfrac{1-e^{-s}}{s(1+e^{-s})}$.

5. Use **Sage** to solve the following problems.
   (a) Find the Laplace transform of $u(t - \pi/4)\cos(t)$.
   (b) Compute the convolution $\sin(t) * \cos(t)$. Do this directly and using the convolution theorem.

6. (a) Use Laplace transforms and partial fractions to solve $y'' + 4y' + 8y = 8$, $y(0) = 2$, $y'(0) = -2$.
   (b) Sketch or plot $f(t) = 1 + \cos(t)u(t - \pi)$ and find $\mathcal{L}[f(t)]$.

7. Use the convolution theorem to find $\mathcal{L}^{-1}\left\{\frac{1}{(s^2+1)^2}\right\}$.

8. Use Laplace transforms to solve the initial value problem

$$y'' + 4y = \delta(t - \pi),\ y(0) = 0,\ y'(0) = -1.$$

   Sketch or plot your solution on the interval $0 \le t \le 2\pi$.

9. Let $f(t) = \begin{cases} 1 & \text{if } 0 \le t \le 3, \\ 0 & \text{if } t \ge 3. \end{cases}$.
   (a) Write $f(t)$ in terms of the unit step function.
   (b) Use (a) to find $\mathcal{L}\{f(t)\}$.
   (c) The charge on an electrical circuit is modeled by the following:

$$q'' + 4q' + 8q = f(t), \qquad q(0) = q'(0) = 0.$$

   Use (b) to solve this initial value problem for the charge.

10. (a) Find
    (i) $\mathcal{L}\{4t^3 - 4e^{-t}\sin(2t)\}$,
    (ii) $\mathcal{L}\{t^2 U(t - 1)\}$,
    (iii) $\mathcal{L}\{t\cos t\}$.
    (b) Find $\mathcal{L}\{f(t)\}$ for the periodic function given over one period by

$$f(t) = \begin{cases} 1, & 0 < t \le 2, \\ 0, & 2 < t < 3. \end{cases}$$

11. (a) Find
    (i) $\mathcal{L}^{-1}\{1/(s^2 + 9) - e^{-3s}/(s - 4)^2\}$,
    (ii) $\mathcal{L}^{-1}\{(s + 4)/(s^2 + 6s + 25)\}$.
    (b) Use the convolution theorem to find $\mathcal{L}^{-1}\{1/((s - 1)^2(s - 3))\}$.

12. The convolution $t * t * t * t * t$ is equal to
    (a) $\frac{1}{s^{10}}$, (b) $\frac{t^9}{9!}$, (c) t^5 , (d) $\frac{1}{(s^2+1)^4}$, (e) none of the above.

Chapter 3

Matrix theory and systems of DEs

> We [Kaplansky and I] share a philosophy about linear algebra: we think basis-free, we write basis-free, but when the chips are down we close the office door and compute with matrices like fury.
>
> —*Paul Halmos*

> ...there is no study in the world which brings into more harmonious action all the faculties of the mind than [mathematics] ...
>
> —*James Joseph Sylvester*

To handle systems of differential equations, in which there is more than one dependent variable, one must learn some linear algebra. It is best to take a full course in that subject, but for those cases where that is not possible we aim to provide the required background in this chapter. In contrast to a linear algebra textbook per se, here we will omit many proofs. Linear algebra is a tremendously useful subject to understand, and we encourage the reader to take a more complete course in it if at all possible. An excellent free reference for linear algebra is the text by Robert Beezer [B-rref].

3.1 Quick survey of linear algebra

3.1.1 Matrix arithmetic

A matrix is a two-dimensional rectangular array of numbers. We always refer to the entry in the ith row and jth column as the (i, j)-*entry*—that is, we always refer to the row index first. A matrix having m rows and n columns is called an $m \times n$ *matrix*. A matrix with $m = n$ is called a *square* matrix.

If A is an $m \times n$ matrix then the *transpose of* A, denoted A^T, is the $n \times m$ matrix where the rows of A^T are the columns of A and the columns of A^T are the rows of A. For example, if

$$A = \begin{pmatrix} 1 & 2 & 3 \\ 4 & 5 & 6 \end{pmatrix}$$

then

$$A^T = \begin{pmatrix} 1 & 4 \\ 2 & 5 \\ 3 & 6 \end{pmatrix}.$$

Two matrices can be added together if they have the same dimensions (the same numbers of rows and columns). The sum is defined to be the "componentwise sum": the matrix whose (i, j) entry is the sum of the (i, j) entries of the summand matrices. For example,

$$\begin{pmatrix} 1 & 0 \\ 0 & 1 \end{pmatrix} + \begin{pmatrix} -1 & 1 \\ 2 & -3 \end{pmatrix} = \begin{pmatrix} 0 & 1 \\ 2 & -2 \end{pmatrix}.$$

Matix multiplication is not defined in the same way. A matrix product $C = AB$ is defined if matrix A has the same number of columns as the number of rows in B. In other words, A must have dimensions (m, p) and matrix B must have dimensions (p, n). If A has *row* vectors A_i $(1 \le i \le m$, each of which is a vector in $\mathbb{R}^p)$ and B has *column* vectors B_j $(1 \le j \le n$, each of which is a vector in $\mathbb{R}^p)$ then the (i, j)th entry of AB is the dot product $A_i \cdot B_j$. Therefore, the resulting matrix C has dimensions (m, n), with the following formula for the (i, j) entry of C (denoted $c_{i,j}$) in terms of the entries of A and B:

$$c_{i,j} = A_i \cdot B_j = \sum_k a_{i,k} b_{k,j}.$$

Lemma 3.1.1. The matrix product is associative: if A, B, and C are matrices and the products AB and BC are defined, then $A(BC) = (AB)C$.

Proof. We can directly verify this property from the definition of the matrix product. The entry of the ith row and jth column of the product $(AB)C$ is

$$\sum_m \left(\sum_k a_{ik} b_{km}\right) c_{mj} = \sum_k \sum_m a_{ik} b_{km} c_{mj}$$
$$= \sum_k a_{ik} \sum_m (b_{km} c_{mj}) = A(BC).$$

□

It is extremely important to note that matrix multiplication is not commutative—it is not always true that $AB = BA$. Here are some examples illustrating this:

$$\begin{pmatrix} 1 & 2 & 3 \\ 4 & 5 & 6 \end{pmatrix} \begin{pmatrix} 1 & 2 \\ 3 & 4 \\ 5 & 6 \end{pmatrix} = \begin{pmatrix} 22 & 28 \\ 49 & 64 \end{pmatrix},$$

$$\begin{pmatrix} 1 & 2 \\ 3 & 4 \\ 5 & 6 \end{pmatrix} \begin{pmatrix} 1 & 2 & 3 \\ 4 & 5 & 6 \end{pmatrix} = \begin{pmatrix} 9 & 12 & 15 \\ 19 & 26 & 33 \\ 29 & 40 & 51 \end{pmatrix}.$$

We say two $n \times n$ matrices A, B are *similar* if there is a matrix C such that

$$AC = CB.$$

Unlike real numbers, in general, two matrices are not similar. For example,

$$A = \begin{pmatrix} 2 & 0 \\ 0 & 3 \end{pmatrix}, \quad B = \begin{pmatrix} 3 & 0 \\ 0 & 2 \end{pmatrix}$$

are similar (why?) but

$$A = \begin{pmatrix} 2 & 0 \\ 0 & 3 \end{pmatrix}, \quad B = \begin{pmatrix} 3 & 0 \\ 0 & 4 \end{pmatrix}$$

are not similar.

The analoges of the number 1 for matrices are the identity matrices. The $n \times n$ *identity matrix* is denoted I_n (or often just I if the context is clear) and is defined as the matrix whose diagonal entries (at positions (i, i) for $i \in (1, \ldots, n)$) are 1 and whose off-diagonal entries are zero:

$$I_n = \begin{pmatrix} 1 & 0 & \ldots & 0 & 0 \\ 0 & 1 & \ldots & 0 & 0 \\ \vdots & & \ddots & & \vdots \\ 0 & 0 & \ldots & 1 & 0 \\ 0 & 0 & \ldots & 0 & 1 \end{pmatrix}.$$

A matrix with all entries equal to zero is called the *zero matrix*. Another important difference between matrix multiplication and multiplication with scalars is that it is possible for the product of two nonzero matrices to equal the zero matrix. For example,

$$\begin{pmatrix} 1 & 2 & 3 \\ 4 & 5 & 6 \end{pmatrix} \begin{pmatrix} 1 & 1 \\ -2 & -2 \\ 1 & 1 \end{pmatrix} = \begin{pmatrix} 0 & 0 \\ 0 & 0 \end{pmatrix}.$$

This also means that one cannot always cancel common factors in a matrix equation—i.e., for matrices A, B, and C, if $AB = AC$ it is not necessarily true that $B = C$. In order to perform such a cancellation, we need the matrix A to have a stronger property than merely being nonzero—it needs to be invertible.

Definition 3.1.1. A $n \times n$ matrix A is *invertible* if there exists a matrix B such that $AB = BA = I_n$. In this case, the *inverse* of A is defined to be $A^{-1} = B$.

For example, the matrix

$$A = \begin{pmatrix} 2 & 0 \\ 0 & 3 \end{pmatrix}$$

is invertible and

$$A = \begin{pmatrix} 1/2 & 0 \\ 0 & 1/3 \end{pmatrix}.$$

We shall discuss matrix inverses more below.

Exercises

1. For $A = \begin{pmatrix} a & b \\ c & d \end{pmatrix}$, compute the product $A^2 = A \cdot A$.

2. For the following three pairs of matrices, compute whatever products between them are defined (i.e., AB or BA or both).

   (a)
   $$A = \begin{pmatrix} 1 & -2 & 1 \end{pmatrix}, \quad B = \begin{pmatrix} 1 & 3 & 0 \\ 1 & 2 & 1 \end{pmatrix}.$$

   (b)
   $$A = \begin{pmatrix} 2 & 0 & 1 & 0 \end{pmatrix}, \quad B = \begin{pmatrix} 1 & 0 \\ 0 & 2 \end{pmatrix}.$$

   (c)
   $$A = \begin{pmatrix} 3 \\ 4 \end{pmatrix}, \quad B = \begin{pmatrix} 1 & 1 \\ 2 & 2 \\ -1 & 1 \end{pmatrix}.$$

3. Find a 2×2 matrix A with all nonzero entries such that $A^2 = 0$.

4. Find a 2×2 matrix A with each diagonal element equal to zero such that $A^2 = -I$. (The existence of such matrices is important, as it lets us represent complex numbers by real matrices.)

5. Find some 2×2 matrices A such that $A^2 = I$. Try to categorize all such matrices. Do you think you have found all of them?

6. Let A be the 4×1 (column) matrix of all 1s and B be the 1×4 (row) matrix $(1, 2, 3, 4)$. Compute

   (a) the matrix product $A \cdot B$;

   (b) the matrix product $B \cdot A$.

7. Compute the inverse of A to find a matrix X such that $AX = B$ if $A = \begin{pmatrix} 5 & 3 \\ 3 & 1 \end{pmatrix}$ and $B = \begin{pmatrix} 3 & 2 & 4 \\ 1 & 6 & -4 \end{pmatrix}$.

8. Verify that if A and B are invertible $n \times n$ matrices, then AB is invertible with inverse $B^{-1}A^{-1}$.

9. Show that if A, B, and C are invertible matrices then $(ABC)^{-1} = C^{-1}B^{-1}A^{-1}$.

3.2 Row reduction and solving systems of equations

Row reduction is the engine that drives a lot of the machinery of matrix theory. What we call *row reduction* others call *computing the reduced row echelon form* or *Gauss-Jordan reduction.* The term *Gauss elimination* is usually reserved for the process of solving a system of equations by first putting them in upper-triangular form, then using substitution to solve for the unknowns.

3.2.1 The Gauss elimination game

This is actually a discussion of solving systems of equations using the method of row reduction, but it's more fun to formulate it in terms of a game.

To be specific, let's focus on a 2×2 system (by "2×2," we mean two equations in the two unknowns x, y):

$$\begin{cases} ax + by = r_1 \\ cx + dy = r_2 \end{cases} \tag{3.1}$$

Here a, b, c, d, r_1, r_2 are given constants. Putting these two equations down together means they should be solved simultaneously, not individually. In geometric terms, you may think of each equation above as a line in the plane. To solve them simultaneously, you are to find the point of intersection (if it exists) of these two lines. Since a, b, c, d, r_1, r_2 have not been specified, it is conceivable that

- no solutions exist (the lines are parallel but distinct);
- infinitely many solutions exist (the lines are the same);
- there is exactly one solution (the lines are distinct and not parallel, as in Figure 3.1).

"Usually" there is exactly one solution. Of course, you can solve this by simply manipulating equations since it is such a low-dimensional system, but the object of this section is to show you a method of solution that is "scalable" to "industrial-sized" problems (say 1000×1000 or larger).

Strategy

Step 1. Write down the *augmented matrix* of (3.1):

$$A = \begin{pmatrix} a & b & r_1 \\ c & d & r_2 \end{pmatrix}.$$

This is simply a matter of stripping off the unknowns and recording the coefficients in an array. Of course, the system must be written in "standard form" (all the terms with "x" get aligned together in a single column, all the terms with y get aligned.) to do this correctly.
Step 2. Play the Gauss elimination game (described below) to computing the row-reduced echelon form of A; call it B.
Step 3. Read off the solution (assuming it exists and is unique) from the rightmost column of B.

The Gauss elimination game
Legal moves. These actually apply to any $m \times n$ matrix A with $m < n$.

1. $R_i \leftrightarrow R_j$: You can swap row i with row j.
2. $cR_i \to R_i$ $(c \neq 0)$: You can replace row i with row i multiplied by any nonzero constant c.
3. $cR_i + R_j \to R_i$ $(c \neq 0)$: You can replace row i with row i multiplied by any nonzero constant c plus row j, $j \neq i$.

Note that move 1 simply corresponds to reordering the system of equations (3.1). Likewise, move 2 simply corresponds to scaling equation i in (3.1). In general, these "legal moves" correspond to algebraic operations you would perform on (3.1)) to solve it. However, there are fewer symbols to push around when the augmented matrix is used. These legal moves are also referred to as *elementary row operations*.

Goal. You *win* the game when you can achieve the following situation. Your goal is to find a sequence of legal moves leading to a matrix B satisfying the following criteria:

1. All rows of B have leading non-zero term equal to 1 (the position where this leading term in B occurs is called a *pivot position*).
2. B contains as many 0s as possible.
3. All entries above and below a pivot position must be 0.
4. The pivot position of the ith row is to the left and above the pivot position of the $(i+1)$st row (therefore, all entries below the diagonal of B are 0).

What we call the "Gauss elimination game" is usually referred to as the *method of row reduction.*

This matrix B constructed above is *unique* (this is a theorem which you can find in any text on elementary matrix theory or linear algebra[1]) and is called the *row-reduced echelon form* of A. The matrix B is sometimes written rref(A).

---

[1]For example, [B-rref] or [H-rref].

Definition 3.2.1. The *rank* of the matrix A is the number of nonzero rows in the matrix rref(A).

Remark 3.2.1.

1. There is at most one pivot position in each row and in each column.
2. Every nonzero row has a pivot position, located at the first nonzero entry of that row.
3. If you and your friend both start out playing this game, it is likely your choice of legal moves will differ. That is to be expected. However, you must get the same result in the end.
4. If you try the row reduction method but are unable to finish, it could be because you are creating nonzero terms in the matrix while "killing" others in a nonsystematic way. Try killing terms only after selecting your pivot position and try to do this in a systematic left-to-right, top-to-bottom fashion.

Now it's time for an example.

Example 3.2.1. Solve

$$\left\{ \begin{array}{rcl} x + 2y & = & 3 \\ 4x + 5y & = & 6 \end{array} \right. \tag{3.2}$$

using row reduction.

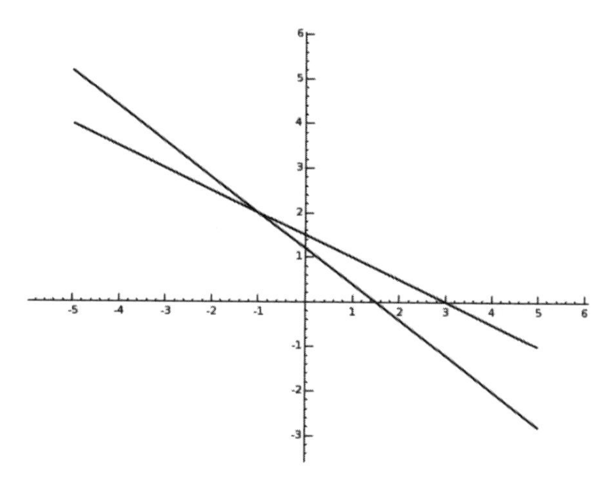

Figure 3.1: Lines $x + 2y = 3$, $4x + 5y = 6$ in the plane.

The augmented matrix is

$$A = \begin{pmatrix} 1 & 2 & 3 \\ 4 & 5 & 6 \end{pmatrix}.$$

One sequence of legal moves is the following:

$-4R_1 + R_2 \to R_2$, which leads to $\begin{pmatrix} 1 & 2 & 3 \\ 0 & -3 & -6 \end{pmatrix}$;

$-(1/3)R_2 \to R_2$, which leads to $\begin{pmatrix} 1 & 2 & 3 \\ 0 & 1 & 2 \end{pmatrix}$;

$-2R_2 + R_1 \to R_1$, which leads to $\begin{pmatrix} 1 & 0 & -1 \\ 0 & 1 & 2 \end{pmatrix}$.

Now we are done (we "won" the game of Gauss elimination), since this matrix satisfies all the goals for a row-reduced echelon form.

The latter matrix corresponds to the system of equations

$$\begin{cases} x + 0y & = & -1, \\ 0x + y & = & 2. \end{cases} \tag{3.3}$$

Since the "legal moves" were simply matrix analogs of algebraic manipulations you would appy to the system (3.2), the solution to (3.2) is the same as the solution to (3.3), which is obviously $x = -1, y = 2$. You can visually check this from the graph in Figure 3.1 .

To find the row-reduced echelon form of

$$\begin{pmatrix} 1 & 2 & 3 \\ 4 & 5 & 6 \end{pmatrix}$$

using **Sage**, just type the following:

Sage

```
sage: MS = MatrixSpace(QQ,2,3)
sage: A = MS([[1,2,3],[4,5,6]])
sage: A
[1 2 3]
[4 5 6]
sage: A.echelon_form()
[ 1  0 -1]
[ 0  1  2]
```

What if there are more variables than equations? Suppose that there are n variables and m linear (simultaneous) equations, with $m < n$. Since each linear equation can cut the number of degrees of freedom down by (at most) 1, the set of solutions must have at least $n - m > 0$ degrees of freedom. In other words, we must expect that there are infinitely many solutions. In general, for solutions to a system $Ax = b$, where A is an $m \times n$ matrix,

$$\text{Number of degrees of freedom of solution} = n - \text{rank}(A).$$

Example 3.2.2. Describe the intersection L of the plane $x+y+z = 1$ with $x+2y+3z = -1$. Note that L is a line in the plane, so we can ask for a parametric representation. We start by computing the row-reduced echelon form of the augmented matrix

$$A = \begin{pmatrix} 1 & 1 & 1 & 1 \\ 1 & 2 & 3 & -1 \end{pmatrix}.$$

The row-reduced echelon form is

$$\mathrm{rref}(A) = \begin{pmatrix} 1 & 0 & -1 & 3 \\ 0 & 1 & 2 & -2 \end{pmatrix},$$

so the resulting system is $x = z + 3$, $y = -2z - 2$. Regarding z as our free variable, this intersection L can be written

$$L = \{(x, y, z) \mid x = z+3, y = -2z-2\} = \{(z+3, -2z-2, z) \mid z \in \mathbb{R}\} = (3, -2, 0)+(1, -2, 1)\cdot\mathbb{R}.$$

The parametric representation of L is

$$x = 3 + t, \quad y = -2 + 2t, \quad z = t,$$

where $t \in \mathbb{R}$.

See §3.2.4 for more higher-dimensional examples.

3.2.2 Solving systems using inverses

There is another method of solving "square" systems of linear equations which we discuss next.

One can rewrite the system (3.1) as a single matrix equation

$$\begin{pmatrix} a & b \\ c & d \end{pmatrix} \begin{pmatrix} x \\ y \end{pmatrix} = \begin{pmatrix} r_1 \\ r_2 \end{pmatrix},$$

or more compactly as

$$A\vec{X} = \vec{r}, \tag{3.4}$$

where $\vec{X} = \begin{pmatrix} x \\ y \end{pmatrix}$ and $\vec{r} = \begin{pmatrix} r_1 \\ r_2 \end{pmatrix}$. How do you solve (3.4)? The obvious thing to do ("divide by A") is the right idea:

$$\begin{pmatrix} x \\ y \end{pmatrix} = \vec{X} = A^{-1}\vec{r}.$$

Here A^{-1} is a matrix with the property that $A^{-1}A = I$, the identity matrix (which satisfies $I\vec{X} = \vec{X}$).

3.2.3 Computing inverses using row reduction

If A^{-1} exists (and it usually does), how do we compute it? There are a few ways. One way is by using a formula. In the 2×2 case, the inverse is given by

$$\begin{pmatrix} a & b \\ c & d \end{pmatrix}^{-1} = \frac{1}{ad - bc} \begin{pmatrix} d & -b \\ -c & a \end{pmatrix}.$$

There is a similar formula for larger sized matrices but it is so unwieldy that it is usually not used to compute the inverse.

Example 3.2.3. In the 2×2 case, the formula above for the inverse is easy to use and we see, for example,

$$\begin{pmatrix} 1 & 2 \\ 4 & 5 \end{pmatrix}^{-1} = \frac{1}{-3}\begin{pmatrix} 5 & -2 \\ -4 & 1 \end{pmatrix} = \begin{pmatrix} -5/3 & 2/3 \\ 4/3 & -1/3 \end{pmatrix}.$$

To find the inverse of

$$\begin{pmatrix} 1 & 2 \\ 4 & 5 \end{pmatrix}$$

using Sage, just type the following:

Sage

```
sage: MS = MatrixSpace(QQ,2,2)
sage: A = MS([[1,2],[4,5]])
sage: A
[1 2]
[4 5]
sage: A^(-1)
[-5/3  2/3]
[ 4/3 -1/3]
```

A better way to compute A^{-1} in general is the following. Compute the row-reduced echelon form of the matrix (A, I), where I is the identity matrix of the same size as A. This new matrix will be (if the inverse exists) (I, A^{-1}). You can read off the inverse matrix from this.

In other words, the following result holds.

Lemma 3.2.1. Let A be an invertible $n \times n$ matrix. The row-reduced echelon form of the $n \times 2n$ matrix (A, I), where I is the identity matrix of the same size as A, is the $n \times 2n$ matrix (I, A^{-1}).

If A is not invertible then the row-reduced echelon form of the $n \times 2n$ matrix (A, I) will not have the form (I, B), for some $n \times n$ matrix B. For example,

$$A = \begin{pmatrix} 1 & 2 \\ 0 & 0 \end{pmatrix}$$

is not invertible, and the row-reduced echelon form of (A, I) cannot be written in the form (I, B), for some 2×2 matrix B.

Here is another illustration.

Example 3.2.4. Solve

$$\begin{cases} x+2y &= 3 \\ 4x+5y &= 6 \end{cases}$$

using (a) row reduction and (b) matrix inverses.

In fact, this system was solved using row reduction in Example 3.2.1 above. The idea was that the result of the Sage commands

```
Sage
sage: MS = MatrixSpace(QQ,2,3)
sage: A = MS([[1,2,3],[4,5,6]])
sage: A
[1 2 3]
[4 5 6]
sage: A.echelon_form()
[ 1  0 -1]
[ 0  1  2]
```

told us that $x = -1$ and $y = 2$. In other words, to get the solution, you just read off the last column of the row-reduced echelon form of the augmented matrix.

The system can be written in matrix form as

$$\begin{pmatrix} 1 & 2 \\ 4 & 5 \end{pmatrix} \begin{pmatrix} x \\ y \end{pmatrix} = \begin{pmatrix} 3 \\ 6 \end{pmatrix},$$

so

$$\begin{pmatrix} x \\ y \end{pmatrix} = \begin{pmatrix} 1 & 2 \\ 4 & 5 \end{pmatrix}^{-1} \begin{pmatrix} 3 \\ 6 \end{pmatrix}.$$

To compute the inverse matrix, so that we can compute x and y, apply the Gauss elimination game to

$$\begin{pmatrix} 1 & 2 & 1 & 0 \\ 4 & 5 & 0 & 1 \end{pmatrix}.$$

Using the same sequence of legal moves as in Example 3.2.1, we get

$-4R_1 + R_2 \to R_2$, which leads to $\begin{pmatrix} 1 & 2 & 1 & 0 \\ 0 & -3 & -4 & 1 \end{pmatrix}$

$-(1/3)R_2 \to R_2$, which leads to $\begin{pmatrix} 1 & 2 & 1 & 0 \\ 0 & 1 & 4/3 & -1/3 \end{pmatrix}$

$-2R_2 + R_1 \to R_1$, which leads to $\begin{pmatrix} 1 & 0 & -5/3 & 2/3 \\ 0 & 1 & 4/3 & -1/3 \end{pmatrix}$.

Therefore the inverse is

$$A^{-1} = \begin{pmatrix} -5/3 & 2/3 \\ 4/3 & -1/3 \end{pmatrix}.$$

Now, to solve the system, compute

$$\begin{pmatrix} x \\ y \end{pmatrix} = \begin{pmatrix} 1 & 2 \\ 4 & 5 \end{pmatrix}^{-1} \begin{pmatrix} 3 \\ 6 \end{pmatrix} = \begin{pmatrix} -5/3 & 2/3 \\ 4/3 & -1/3 \end{pmatrix} \begin{pmatrix} 3 \\ 6 \end{pmatrix} = \begin{pmatrix} -1 \\ 2 \end{pmatrix}.$$

To make Sage do the above computation, just type the following:

Sage

```
sage: MS = MatrixSpace(QQ,2,2)
sage: A = MS([[1,2],[4,5]])
sage: V = VectorSpace(QQ,2)
sage: v = V([3,6])
sage: A^(-1)*v
 (-1, 2)
```

Of course, this again tells us that $x = -1$ and $y = 2$ is the solution to the original system.

We use row reduction to compute the matrix inverse in the next example.

Example 3.2.5. To invert

$$A = \begin{pmatrix} 1 & 2 & 3 \\ 1 & 2 & 2 \\ -1 & -1 & 1 \end{pmatrix},$$

we first augment it with the 3×3 identity matrix, and then row-reduce until we obtain the identity matrix:

$$\begin{pmatrix} 1 & 2 & 3 & 1 & 0 & 0 \\ 1 & 2 & 2 & 0 & 1 & 0 \\ -1 & -1 & 1 & 0 & 0 & 1 \end{pmatrix} \xrightarrow{-R_1+R_2} \begin{pmatrix} 1 & 2 & 3 & 1 & 0 & 0 \\ 0 & 0 & -1 & -1 & 1 & 0 \\ -1 & -1 & 1 & 0 & 0 & 1 \end{pmatrix}$$

$$\xrightarrow{R_1+R_3} \begin{pmatrix} 1 & 2 & 3 & 1 & 0 & 0 \\ 0 & 0 & -1 & -1 & 1 & 0 \\ 0 & 1 & 4 & 1 & 0 & 1 \end{pmatrix} \xrightarrow{Swap(R_2,R_3)} \begin{pmatrix} 1 & 0 & 0 & 4 & -5 & -2 \\ 0 & 1 & 0 & -3 & 4 & 1 \\ 0 & 0 & 1 & 1 & -1 & 0 \end{pmatrix},$$

where the last step also involved a rescaling by a sign: $R_3 \to -R_3$. Reading off the last 3×3 block, we therefore have $A^{-1} = \begin{pmatrix} 4 & -5 & -2 \\ -3 & 4 & 1 \\ 1 & -1 & 0 \end{pmatrix}$.

There are many other applications of row reduction: We can use it to compute a basis for row spaces and column spaces, matrix kernels (null spaces), and orthogonal complements. These are interesting and useful topics, but fall outside a textbook on differential equations. We refer to Beezer's text [B-rref], for details.

Exercises

1. Using Sage, solve

$$\begin{cases} x + 2y + z &= 1, \\ -x + 2y - z &= 2, \\ y + 2z &= 3 \end{cases}$$

using (a) row reduction and (b) matrix inverses (i.e., convert to a matrix equation $A\vec{v} = \vec{b}$; compute $\vec{v} = A^{-1}\vec{b}$).

2. Solve

$$\begin{cases} x + y - z = 1, \\ x - y + z = 2, \\ -x + y + z = 3, \end{cases}$$

using (a) row reduction and (b) matrix inverses.

3. Determine the intersection of planes

$$\begin{cases} x + y - z = 1, \\ x - y + z = 2, \end{cases}$$

as a line in $\mathbb{R}^3$.

4. Solve the system

$$x - y - z = 1,$$

$$-x + y - z = 2,$$

$$x + 2y + 3z = 3,$$

using row reduction of the augmented matrix.

5. Find the inverse A^{-1} of the following matrix by augmenting with the identity and computing the reduced echelon form.

$$A = \begin{pmatrix} 5 & 3 & 6 \\ 4 & 2 & 3 \\ 3 & 2 & 5 \end{pmatrix}.$$

6. Find the inverse A^{-1} of the matrix $A = \begin{pmatrix} 0 & 0 & 1 \\ 2 & 0 & 0 \\ 0 & 1 & 2 \end{pmatrix}$ by using elementary row operations on the matrix A adjoined by the 3×3 identity matrix.

7. (a) Using row reduction, solve

$$\begin{aligned} x+y&=1,\\ y+z&=-1,\\ x+z&=1. \end{aligned}$$

Write down the augmented matrix and its row-reduced echelon form.

(b) Let A be the matrix of coefficients of the system in (a) (so the system can be written as a matrix equation $A\vec{x}=\vec{b}$). Compute the inverse of A.

(c) Compute the 3×3 matrix A^3-3A^2+3A.

8. Let

$$A=\begin{pmatrix} 1 & 1 & 1 & 1\\ -1 & -1 & 1 & -2\\ 2 & 2 & 0 & 1 \end{pmatrix}.$$

Find the row-reduced echelon form of A. Explain your result geometrically (i.e., describe the appearance of the intersection of the planes $x+y+z=1$, $-x-y+z=-2$, and $2x+2y=1$).

9. Solve

$$\begin{cases} 2x+3y=4,\\ 5x+6y=7, \end{cases}$$

using the row-reduced echelon form of the augmented matrix. Label all elementary row operation steps of the row reduction process.

10. Let

$$A=\begin{pmatrix} 0 & 1 & 1 & 1 & -1\\ 1 & 0 & 1 & -2 & 0\\ 1 & 1 & 0 & 1 & 1 \end{pmatrix}.$$

Find the row-reduced echelon form of A.

11. Compute the inverse of

$$A=\begin{pmatrix} 0 & 1 & 1\\ 1 & 0 & 1\\ 1 & 1 & 0 \end{pmatrix},$$

using row reduction.

3.2.4 Solving higher-dimensional linear systems

This is the Gauss-Jordan reduction, revisited.

Example 3.2.6. Solve the linear system

$$\begin{aligned} x + 2y + 3z &= 0, \\ 4x + 5y + 6z &= 3, \\ 7x + 8y + 9z &= 6. \end{aligned}$$

We form the augmented coefficient matrix and row-reduce until we obtain the reduced-row echelon form:

$$\begin{pmatrix} 1 & 2 & 3 & 0 \\ 4 & 5 & 6 & 3 \\ 7 & 8 & 9 & 6 \end{pmatrix} \xrightarrow{-4R_1+R_2} \begin{pmatrix} 1 & 2 & 3 & 0 \\ 0 & -3 & -6 & 3 \\ 7 & 8 & 9 & 6 \end{pmatrix}$$

$$\xrightarrow{-7R_1+R_3} \begin{pmatrix} 1 & 2 & 3 & 0 \\ 0 & -3 & -6 & 3 \\ 0 & -6 & -12 & 6 \end{pmatrix} \xrightarrow{-R_2/3} \begin{pmatrix} 1 & 2 & 3 & 0 \\ 0 & 1 & 2 & -1 \\ 0 & -6 & -12 & 6 \end{pmatrix}$$

$$\xrightarrow{-R_3/6} \begin{pmatrix} 1 & 2 & 3 & 0 \\ 0 & 1 & 2 & -1 \\ 0 & 1 & 2 & -1 \end{pmatrix} \xrightarrow{-R_2+R_3} \begin{pmatrix} 1 & 2 & 3 & 0 \\ 0 & 1 & 2 & -1 \\ 0 & 0 & 0 & 0 \end{pmatrix}$$

$$\xrightarrow{-2R_2+R_1} \begin{pmatrix} 1 & 0 & -1 & 2 \\ 0 & 1 & 2 & -1 \\ 0 & 0 & 0 & 0 \end{pmatrix}$$

Reinterpreting these rows as equations, we have $x - z = 2$ and $y + 2z = -1$. The pivot variables are x and y, and z is the only free variable. Finally, we write the solutions in parametric vector form:

$$\begin{pmatrix} x \\ y \\ z \end{pmatrix} = \begin{pmatrix} 2+t \\ -1-2t \\ t \end{pmatrix} = \begin{pmatrix} 2 \\ -1 \\ 0 \end{pmatrix} + \begin{pmatrix} 1 \\ -2 \\ 1 \end{pmatrix} t.$$

Exercises

1. Reduce the following matrix to reduced echelon form.

$$\begin{pmatrix} 1 & 3 & -6 \\ 2 & 6 & 7 \\ 3 & 9 & 1 \end{pmatrix}.$$

2. Write the following system as a matrix equation, and find all the solutions.

$$\begin{aligned} 3x + 2y + z &= 2, \\ x - y + 2z &= 2, \\ 3x + 7y - 4z &= -2. \end{aligned}$$

3. Use elementary row operations to transform the augmented coefficient matrix of the system below into echelon or reduced echelon form. Use this to write down all the solutions to the system.

$$\begin{aligned} x + 2y + -2z &= 3, \\ 2x + 4y - 2z &= 2, \\ x + 3y + 2z &= -1. \end{aligned}$$

4. Do the same as in Exercise 3, for the system

$$\begin{aligned} x + 2y - z &= -1, \\ x + 3y - 2z &= 0, \\ -x - z &= 3. \end{aligned}$$

5. Under what conditions on a, b, and c does the following system have a unique solution, infinitely many solutions, or no solutions?

$$\begin{aligned} 2x - y + 3z &= a, \\ x + 2y + z &= b, \\ 6x + 2y + 8z &= c. \end{aligned}$$

6. Under what condition (if any) on a does the following system have no solution?

$$\begin{aligned} x + 2y + 3z &= 1, \\ 4x + 5y + 6z &= 1, \\ 7x + 8y + 9z &= a. \end{aligned}$$

7. The system $x + y = 1$, $y - z = 1$, $x + z = 1$ has

   (a) a unique solution, (b) no solutions, (c) infinitely many solutions,
   (d) exactly two solutions, (e) none of these.

3.2.5 Determinants

Square matrices ($n \times n$) have a special number associated with them called the determinant. The determinant is defined inductively, in terms of determinants of smaller matrices. Eventually this induction involves the determinants of small matrices, for which we can define the determinant directly:

- If $n = 1$: $\det(A) = A$ (in this case the matrix A is simply a real number).

- If $n = 2$:

$$\det \begin{pmatrix} a & b \\ c & d \end{pmatrix} = ad - bc.$$

Definition 3.2.2. The *(i, j)th minor* of a matrix A is the determinant of the submatrix obtained by deleting row i and column j from A.

Example 3.2.7. The $(2,2)$-minor of $\begin{pmatrix} 1 & 2 & -1 \\ 0 & 0 & 0 \\ 2 & 3 & 4 \end{pmatrix}$ is equal to $\det \begin{pmatrix} 1 & -1 \\ 2 & 4 \end{pmatrix} = 6$. This is easily verified using Sage:

Sage

```
sage: A = matrix(QQ,[[1,2,-1],[0,0,0],[2,3,4]])
sage: A.matrix_from_rows_and_columns([0,2],[0,2])
[ 1 -1]
[ 2  4]
sage: det(A.matrix_from_rows_and_columns([0,2],[0,2]))
6
sage: A.minors(2)[4]
6
```

The last command might be a little mysterious. First, Sage follows Python by starting its counting at 0 instead of 1. The command `A.minors(2)` returns an ordered list of all 2×2 minors of A. The ordering is in "lexicographical row major ordering," meaning you start in the upper left-hand corner for the first entry, and work your way left to right, top to bottom, to the last entry. The $(2,2)$ minor is actually the 5th entry in this list if you start counting at 1. We want the 4th entry, as in the command above.

Let A be an $n \times n$ matrix, with $n > 1$, and let A_{ij} denote the $(n-1) \times (n-1)$ submatrix obtained by removing the ith row and jth column of A (so $\det(A_{ij})$ is the (i,j)-minor of A), $1 \leq i, j \leq n$. Then the *determinant of* A, denoted $\det(A)$, is defined inductively for $n > 1$ by

$$\det(A) = a_{11}\det(A_{11}) - a_{12}\det(A_{12}) + \cdots + (-1)^{n-1}a_{1n}\det(A_{1n}) = \sum_{j=1}^{n}(-1)^{j+1}a_{1j}\det(A_{1j}).$$

This is called the *Laplace expansion across the first row.*

There is an analogous Laplace expansion across any row (e.g., the kth row):

$$\begin{aligned} \det(A) &= (-1)^{k-1}(a_{k1}\det(A_{k1}) - a_{k2}\det(A_{k2}) + \cdots + (-1)^{n-1}a_{kn}\det(A_{kn})) \\ &= \textstyle\sum_{j=1}^{n}(-1)^{j+k}a_{kj}\det(A_{kj}), \end{aligned} \quad (3.5)$$

and down any column (e.g., the kth column):

$$\begin{aligned}\det(A) &= (-1)^{k-1}(a_{1k}\det(A_{1k}) - a_{2k}\det(A_{2k}) + \cdots + (-1)^{n-1}a_{nk}\det(A_{nk})) \\ &= \sum_{j=1}^{n}(-1)^{j+k}a_{jk}\det(A_{jk}).\end{aligned} \tag{3.6}$$

It is a theorem that these different formulas (3.5) and (3.6) for the determinant yield the same number (for details, see the textbook on linear algebra by Beezer [B-rref]).

Geometric interpretation

If $A = \begin{pmatrix} a & b \\ c & d \end{pmatrix}$, the absolute value of the determinant $|\det A|$ can be viewed as the area of the parallelogram with vertices at $(0,0)$, (a,b), $(a+c, b+d)$, and (c,d).

Exercises

1. Compute the determinant of $A = \begin{pmatrix} 2 & 3 & 4 & 5 \\ 1 & 0 & 0 & 2 \\ 0 & 1 & 0 & -1 \\ 0 & 1 & 0 & 1 \end{pmatrix}$.

2. Compute the determinant of $A = \begin{pmatrix} 3 & 1 & 0 & 2 \\ 0 & 1 & 20 & 2 \\ 0 & 0 & 1 & -63 \\ 9 & 3 & 0 & 1 \end{pmatrix}$.

3. Show that if a matrix A has the property $A^2 = A$, then either $\det(A) = 0$ or $\det(A) = 1$.

4. A matrix is called *orthogonal* if $A^T = A^{-1}$. Show that if A is orthogonal then $\det(A) = \pm 1$.

5. The $n \times n$ Vandermonde determinant is defined as

$$V(x_1, x_2, \dots, x_n) = \begin{vmatrix} 1 & x_1 & x_1^2 & \cdots & x_1^{n-1} \\ 1 & x_2 & x_2^2 & \cdots & x_2^{n-1} \\ \vdots & \vdots & \vdots & \ddots & \vdots \\ 1 & x_n & x_n^2 & \cdots & x_n^{n-1} \end{vmatrix}.$$

   Show that

   (a) the 2×2 Vandermonde determinant $V(a,b) = b - a$;

   (b) the 3×3 Vandermonde determinant $V(1,a,b)$ can be factored into $(a-1)(b-1)(b-a)$.

6. (*) Find a 3×3 matrix with strictly positive entries ($a_{ij} > 0$ for each entry a_{ij}) whose determinant is equal to 1. Try to find such a matrix with the smallest sum of entries, that is, such that $\sum_{i=1}^{3}\sum_{i=1}^{3} a_{ij} = a_{11} + a_{12} + \cdots + a_{33}$ is as low as possible.

3.2.6 Elementary matrices and computation of determinants

Determinants are extremely computationally expensive to compute in general. If you use the inductive definition directly, the number of required operations (additions and multiplications) is a rapidly increasing function of the size of the $(n \times n)$ matrix—greater than $n!$ as soon as $n > 1$.

One way to dramatically decrease the effort required to compute determinants involves the relation of an elementary row operation to the determinant of the original and resulting matrices. Each elementary row operation can be expressed as a matrix multiplication; the matrices corresponding to each operation are called *elementary matrices.* Another way to describe the elementary matrices is through this definition:

Definition 3.2.3. A $n \times n$ matrix is an *elementary matrix* if it can be obtained from the action of a single elementary row operation upon the identity matrix I_n.

Each elementary matrix corresponds to the elementary row operation that produces it from the identity. Here are some examples of elementary matrices:

$$\begin{pmatrix} 1 & 0 & 0 \\ 0 & 1 & 0 \\ 0 & 0 & 1 \end{pmatrix} \xrightarrow{-R_1+R_2} \begin{pmatrix} 1 & 0 & 0 \\ -2 & 1 & 0 \\ 0 & 0 & 1 \end{pmatrix} = E_1,$$

$$\begin{pmatrix} 1 & 0 & 0 \\ 0 & 1 & 0 \\ 0 & 0 & 1 \end{pmatrix} \xrightarrow{5R_3} \begin{pmatrix} 1 & 0 & 0 \\ 0 & 1 & 0 \\ 0 & 0 & 5 \end{pmatrix} = E_2,$$

$$\begin{pmatrix} 1 & 0 & 0 \\ 0 & 1 & 0 \\ 0 & 0 & 1 \end{pmatrix} \xrightarrow{Swap(R_2,R_3)} \begin{pmatrix} 1 & 0 & 0 \\ 0 & 0 & 1 \\ 0 & 1 & 0 \end{pmatrix} = E_3.$$

Because of all the zeros in E_1, E_2, and E_3, it is not hard to compute their determinants: $\det(E_1) = 1$, $\det(E_2) = 5$, and $\det(E_3) = -1$. From these examples you might guess the following theorem.

Theorem 3.2.1. The determinants of the elementary matrices corresponding to the three types of elementary row operations are

- $\det(E) = -1$ for swapping row i with row j.
- $\det(E) = c$ for replacing row i with row i multiplied by any non-zero constant c.
- $\det(E) = 1$ for replacing row i with row i multiplied by any non-zero constant c plus row j, $j \neq i$.

This is proved in most linear algebra books, such as [B-rref].

The most important case in the above theorem is the last one—it means that the determinant of a matrix is unaffected by adding a multiple of one row to another (since $\det(A) = \det(E)\det(A) = \det(EA)$ where E is the corresponding elementary matrix for each step). This can be used to partially row-reduce a matrix A into a form in which the determinant is easy to compute. If the matrix can be made upper- or lower-triangular (all entries below or above the diagonal are zero), then the determinant is simply the product of the diagonal entries.

Example 3.2.8. We will compute the determinant of

$$A = \begin{pmatrix} 2 & 1 & -3 & 2 \\ 0 & 1 & 4 & 3 \\ 0 & 0 & 1 & 1 \\ 4 & 1 & -6 & 1 \end{pmatrix}$$

by exploiting elementary row operations which add a multiple of one row to another.

The initial matrix is almost upper-triangular, so we will reduce the nonzero entries below the diagonal to zero:

$$\begin{pmatrix} 2 & 1 & -3 & 2 \\ 0 & 1 & 4 & 3 \\ 0 & 0 & 1 & 1 \\ 4 & 1 & -6 & 1 \end{pmatrix} \xrightarrow{-2R_1+R_4} \begin{pmatrix} 2 & 1 & -3 & 2 \\ 0 & 1 & 4 & 3 \\ 0 & 0 & 1 & 1 \\ 0 & -1 & 0 & -3 \end{pmatrix},$$

$$\begin{pmatrix} 2 & 1 & -3 & 2 \\ 0 & 1 & 4 & 3 \\ 0 & 0 & 1 & 1 \\ 0 & -1 & 0 & -3 \end{pmatrix} \xrightarrow{R_2+R_4} \begin{pmatrix} 2 & 1 & -3 & 2 \\ 0 & 1 & 4 & 3 \\ 0 & 0 & 1 & 1 \\ 0 & 0 & 4 & 0 \end{pmatrix},$$

$$\begin{pmatrix} 2 & 1 & -3 & 2 \\ 0 & 1 & 4 & 3 \\ 0 & 0 & 1 & 1 \\ 0 & 0 & 4 & 0 \end{pmatrix} \xrightarrow{-4R_3+R_4} \begin{pmatrix} 2 & 1 & -3 & 2 \\ 0 & 1 & 4 & 3 \\ 0 & 0 & 1 & 1 \\ 0 & 0 & 0 & -4 \end{pmatrix}.$$

Now we can compute $\det(A) = 2(-4) = -8$ by multiplying the diagonal elements.

Exercises

1. Compute the determinant of

$$A = \begin{pmatrix} 3 & 1 & 0 & 2 \\ 0 & 1 & 20 & 2 \\ 0 & 0 & 1 & -63 \\ 9 & 3 & 0 & 1 \end{pmatrix}$$

using elementary row operations.

2. Compute the determinant of

$$\begin{pmatrix} 2 & 2 & 2 & 2 & 2 \\ 2 & 3 & 3 & 3 & 3 \\ 2 & 3 & 4 & 4 & 4 \\ 2 & 3 & 4 & 5 & 5 \\ 2 & 3 & 4 & 5 & 4 \end{pmatrix}$$

using elementary row operations.

3.2.7 Vector spaces

So far we have worked with vectors which are n-tuples of numbers. But the essential operations we perform on such tuples can be generalized to a much wider class of objects, which we also call vectors. A collection of vectors that are compatible in the sense that they can be added together and multiplied by some sort of scalar is called a vector space. Here is a more formal definition.

Definition 3.2.4. A *vector space* is a set S on which there is an operation of addition which take pairs of elements of S to elements of S, together with a field[2] of numbers K for which there is an operation $\cdot$ of scalar multiplication. These operations must have the following properties:

1. There exists a unique element $\mathbf{0} \in S$ (called the *zero vector*) such that $\mathbf{0} + v = v$ for all $v \in S$.
2. $v + w = w + v$ for all $v, w \in S$.
3. $v + (w + x) = (v + w) + x$ for all $v, w, x \in S$.
4. $(-1) \cdot v + v = \mathbf{0}$ for all $v \in S$.
5. $0 \cdot v = \mathbf{0}$.
6. $1 \cdot v = v$ for all $v \in S$.
7. $(s + t) \cdot (v + w) = s \cdot v + t \cdot v + s \cdot w + t \cdot w$ for all $s, t \in K$ and all $v, w \in S$.
8. $s \cdot (t \cdot v) = (s \cdot t) \cdot v$ for all $s, t \in K$ and all $v \in S$.

For convenience, we do not usually indicate scalar multiplication with a $\cdot$ but by simple juxtaposition. For example, we often write "5 times the vector v" simply as $5v$, instead of $5 \cdot v$, following the usual convention.

For our purposes the field K will always be either the field of real numbers (denoted $\mathbb{R}$) or the field of complex numbers (denoted by $\mathbb{C}$). There are many other fields of numbers used in mathematics but they will not be addressed here so we will not formally define the concept of a field. Unless indicated otherwise, we will use $K = \mathbb{R}$.

Here are some vector space examples:

[2] A field will not be defined here. We shall always assume that the field is either the reals $\mathbb{R}$ or the complexes $\mathbb{C}$.

1. For each positive integer n, the set of lists of n real numbers, $\mathbb{R}^n$, is a vector space. For $n = 2$ and $n = 3$ these are the familiar real plane vectors and real 3-space vectors.

2. The set of real-valued polynomial functions forms a vector space. The addition is the usual addition of functions.

3. For each positive integer n, the set of polynomials of degree at most n forms a vector space. We will denote this space by P_n.

Definition 3.2.5. A *subspace* of a vector space V is a subset of elements of V that is itself a vector space.

The important thing to understand about subspaces is that they must be closed under the operations of scalar multiplication and vector addition—not every subset of a vector space is a subspace. In particular, every subspace W must contain the 0 vector since if $w \in W$, then $-w \in W$, and then so is $w + -w = 0$.

Here are some vector subspace examples:

1. The set $W = \{(x, x) \mid x \in \mathbb{R}\}$ is a subspace of $\mathbb{R}^2$. If we think of $\mathbb{R}^2$ as the x, y plane, then W is simply the line $y = x$.

2. The set $W = \{(x, -2x) \mid x \in \mathbb{R}\}$ is a subspace of $\mathbb{R}^2$. If we think of $\mathbb{R}^2$ as the x, y plane, then W is simply the line $y = -2x$. In fact every line through the origin is a subspace.

Exercises

1. Are the following sets subspaces of $\mathbb{R}^3$ or not? If not, explain why.

   (a) $\{(x_1, x_2, x_3) \mid x_1 + x_2 + x_3 = 0\}$,
   (b) $\{(x_1, x_2, x_3) \mid x_1 x_2 x_3 = 0\}$,
   (c) $\{(x_1, x_2, 0) \mid x_1 = 5x_2\}$,
   (d) The span of the vectors $(1, 2, 3)$, $(4, 5, 6)$ and $(7, 8, 9)$.

2. If W is the subset of all vectors (x, y) in $\mathbb{R}^2$ such that $|x| = |y|$, is W a vector subspace or not?

3. Is the set of all real-valued functions of one variable with the property that $f(0) = 0$ a vector space?

4. Is the set of all real-valued functions of one variable with the property that $f(x) = f(-x)$ for all x a vector space?

5. Is the set of all real-valued functions of one variable with the property that $f(x) = f(-x) + 1$ for all x a vector space?

3.2.8 Bases, dimension, linear independence, and span

This section briefly recalls some fundamental notions of linear algebra.

Definition 3.2.6. A set of nonzero vectors $\{v_1, \ldots, v_m\}$ is *linearly dependent* if there are constants $c_1, \ldots, c_m$ which are not all zero for which $c_1v_1 + \cdots + c_mv_m = 0$. If $\{v_1, \ldots, v_m\}$ is not linearly dependent then we say it is *linearly independent* .

Recall, that the *rank* of an $m \times n$ matrix A is the number of nonzero rows in the row-reduced echelon form of A. It turns out that this is also the number of rows of A (considered as row vectors in $\mathbb{R}^n$) which are linearly independent.

An expression of the form $c_1v_1 + \cdots + c_mv_m$, for constants $c_1, \ldots, c_m$, is called a *span* or *linear combination* of the vectors in $\{v_1, \ldots, v_m\}$. The *row span* of an $m \times n$ matrix A is the vector space spanned by the rows of A (considered as row vectors in $\mathbb{R}^n$). The *column span* of an $m \times n$ matrix A is the vector space spanned by the columns of A (considered as column vectors in $\mathbb{R}^m$).

Example 3.2.9. The vectors $v_1 = \begin{pmatrix} 1 \\ 2 \\ 3 \end{pmatrix}$, $v_2 = \begin{pmatrix} 4 \\ 5 \\ 6 \end{pmatrix}$, and $v_3 = \begin{pmatrix} 7 \\ 8 \\ 9 \end{pmatrix}$ are linearly dependent. To see this, we can write the linear dependence condition $c_1v_1 + c_2v_2 + c_3v_3 = 0$ as a matrix-vector equation:

$$\begin{pmatrix} 1 & 4 & 7 \\ 2 & 5 & 8 \\ 3 & 6 & 9 \end{pmatrix} \begin{pmatrix} c_1 \\ c_2 \\ c_3 \end{pmatrix} = \begin{pmatrix} 0 \\ 0 \\ 0 \end{pmatrix}$$

and solve the system with row reduction. Since it is a homogeneous system (i.e. the left-hand side is the zero vector), we only need to reduce the coefficient matrix:

$$\begin{pmatrix} 1 & 4 & 7 \\ 2 & 5 & 8 \\ 3 & 6 & 9 \end{pmatrix} \xrightarrow{-2R_1+R_2} \begin{pmatrix} 1 & 4 & 7 \\ 0 & -3 & -6 \\ 3 & 6 & 9 \end{pmatrix} \xrightarrow{-3R_1+R_3} \begin{pmatrix} 1 & 4 & 7 \\ 0 & -3 & -6 \\ 0 & -6 & -12 \end{pmatrix}$$

$$\xrightarrow{-2R_2+R_3} \begin{pmatrix} 1 & 4 & 7 \\ 0 & -3 & -6 \\ 0 & 0 & 0 \end{pmatrix} \xrightarrow{-R_2/3} \begin{pmatrix} 1 & 4 & 7 \\ 0 & 1 & 3 \\ 0 & 0 & 0 \end{pmatrix} \xrightarrow{-4R_2+R_1} \begin{pmatrix} 1 & 0 & -5 \\ 0 & 1 & 3 \\ 0 & 0 & 0 \end{pmatrix},$$

which tells us that c_3 is a free variable, so it can be chosen to be nonzero, and the vectors are linearly dependent. As a particular example, we could choose $c_3 = 1$, and then $c_1 = 5$ and $c_2 = -3$, so $5v_1 - 3v_2 + v_3 = 0$.

The above example illustrates one way of determining if a set of vectors is linearly dependent or not. The following theorem generalizes the method used in the above example.

Theorem 3.2.2. Let $S = \{v_1, \ldots, v_k\}$ be k column vectors in $\mathbb{R}^n$. Let $A = (v_1, \ldots, v_k)$ denote the $n \times k$ matrix whose columns are the v_i's. The set S is linearly independent if and only if the matrix $\mathrm{rref}(A^T)$ has no "all 0s" row(s) in it.

Definition 3.2.7. A *basis* of a vector space V is a set of linearly independent vectors that span V.

Example 3.2.10. The set of vectors $\{(1,-1,0),(0,1,-1),(1,1,1)\}$ is a basis for $\mathbb{R}^3$. To see that they span $\mathbb{R}^3$, we can find explicit solutions to the system

$$\begin{pmatrix} 1 & 0 & 1 \\ -1 & 1 & 1 \\ 0 & -1 & 1 \end{pmatrix} \begin{pmatrix} c_1 \\ c_2 \\ c_3 \end{pmatrix} = \begin{pmatrix} x \\ y \\ z \end{pmatrix},$$

for any element $(x,y,z) \in \mathbb{R}^3$, by row-reducing the coefficient matrix (whose columns are the vectors of the basis). We find that $c_1 = \frac{2}{3}x - \frac{1}{3}y - \frac{1}{3}z$, $c_2 = \frac{1}{3}x + \frac{1}{3}y - \frac{2}{3}z$, and $c_3 = \frac{1}{3}x + \frac{1}{3}y + \frac{1}{3}z$.

Since $c_1 = c_2 = c_3 = 0$ is the unique solution for $(x,y,z) = (0,0,0)$, the set of vectors is linearly independent.

Theorem 3.2.3. Every basis of a vector space has the same number of elements, if finite.

Definition 3.2.8. The *dimension* of a vector space V is the number of elements in a basis of V. If the bases of V are not finite, we say V is infinite dimensional.

The following theorem can be found in any textbook on linear algebra (for example, Beezer [B-rref]).

Theorem 3.2.4. Let $S = \{v_1, \ldots, v_k\}$ be k column vectors in $\mathbb{R}^n$. Let V denote the span of S and let $A = (v_1, \ldots, v_k)$ denote the $n \times k$ matrix whose columns are the v_i's. The dimension of the vector space V is the rank of the matrix A. This is the number of nonzero rows in $\mathrm{rref}(A^T)$.

Example 3.2.11. Compute the dimension of the subspace W of $\mathbb{R}^4$ spanned by the vectors $v_1 = (1,1,1,1)^T$, $v_2 = (1,2,1,2)^T$, and $v_3 = (1,3,1,3)^T$.

We can use Theorem 3.2.4, which says it is the number of nonzero rows in

$$\mathrm{rref}\begin{pmatrix} 1 & 1 & 1 & 1 \\ 1 & 2 & 1 & 2 \\ 1 & 3 & 1 & 3 \end{pmatrix} = \begin{pmatrix} 1 & 0 & 1 & 0 \\ 0 & 1 & 0 & 1 \\ 0 & 0 & 0 & 0 \end{pmatrix}.$$

To compute the dimension, we need to know if v_1, v_2, and v_3 form a basis of W. Since W is defined as their span, we need to check only if they are linearly dependent or not. As before, we do this by row-reducing a matrix whose columns consist of the v_i:

$$\begin{pmatrix} 1 & 1 & 1 \\ 1 & 2 & 3 \\ 1 & 1 & 1 \\ 1 & 2 & 3 \end{pmatrix} \xrightarrow{-R_1+R_2} \begin{pmatrix} 1 & 1 & 1 \\ 0 & 1 & 2 \\ 1 & 1 & 1 \\ 1 & 2 & 3 \end{pmatrix} \xrightarrow{-R_1+R_3} \begin{pmatrix} 1 & 1 & 1 \\ 0 & 1 & 2 \\ 0 & 0 & 0 \\ 1 & 2 & 3 \end{pmatrix}$$

$$\xrightarrow{-R_1+R_4} \begin{pmatrix} 1 & 1 & 1 \\ 0 & 1 & 2 \\ 0 & 0 & 0 \\ 0 & 1 & 2 \end{pmatrix} \xrightarrow{-R_2+R_4} \begin{pmatrix} 1 & 1 & 1 \\ 0 & 1 & 2 \\ 0 & 0 & 0 \\ 0 & 0 & 0 \end{pmatrix}.$$

This shows that only two of the three vectors are linearly independent, so the dimension of W is equal to 2.

Exercises

1. Find a basis for the subspace defined by the following equations for $(x_1, x_2, x_3, x_4, x_5) \in \mathbb{R}^5$:

$$\begin{aligned} 2x_1 + x_3 - 2x_4 - 2x_5 &= 0, \\ x_1 + 2x_3 - x_4 + 2x_5 &= 0, \\ -3x_1 - 4x_3 + 3x_4 - 2x_5 &= 0. \end{aligned}$$

2. Consider the triple of vectors $v_1 = (0, 2, 3, -2)$, $v_2 = (3, -1, 4, 1)$, and $v_3 = (6, -8, -1, 8)$.

   (a) Is the set $\{v_1, v_2, v_3\}$ linearly independent or dependent?

   (b) What is the dimension of their span?

   (c) If the vectors are linearly independent, find an additional vector v_4 that makes $\{v_1, v_2, v_3, v_4\}$ a basis for $\mathbb{R}^4$. If they are linearly dependent, write v_1 as a linear combination of v_2 and v_3.

3. Find a basis for the subspace of $\mathbb{R}^3$ given by $x - 2y + 7z = 0$.

4. Find a basis for the subspace of $\mathbb{R}^3$ given by $x = z$.

5. Find a basis for the subspace of all vectors (x_1, x_2, x_3, x_4) in $\mathbb{R}^4$ such that $x_1 + x_2 = x_3 + x_4$.

6. Let

$$A = \begin{pmatrix} 1 & 2 & 3 & 4 \\ 5 & 6 & 7 & 8 \\ 9 & 10 & 11 & 12 \\ 13 & 14 & 15 & 16 \end{pmatrix}.$$

   Let V be the vector space spanned by its columns. Find a basis for V.

7. Let A be as in Exercise 6. Solve the same problem, but let V be the vector space spanned by its rows.

8. Let

$$A = \begin{pmatrix} 1 & 2 & 3 & 4 \\ 5 & 6 & 2 & -4 \\ 3 & 4 & -2 & 1 \end{pmatrix}.$$

Find the rank of A.

9. Let

$$A = \begin{pmatrix} 1 & 2 & 3 & 4 \\ 0 & 1 & 2 & -4 \\ 3 & 4 & -2 & 1 \end{pmatrix}.$$

Find a basis for the column span.

10. Let

$$A = \begin{pmatrix} 1 & 2 & 3 & -3 \\ 5 & 0 & 2 & -4 \\ 3 & 1 & -2 & 1 \end{pmatrix}.$$

Find a basis for the row span.

11. Determine if the rows of the matrix A form a linearly independent set, where

$$A = \begin{pmatrix} 1 & 2 & 3 & 4 \\ 0 & 1 & 2 & -4 \\ 3 & 4 & -2 & 1 \end{pmatrix}.$$

12. Determine if the columns of the matrix A form a linearly independent set, where

$$A = \begin{pmatrix} 1 & 2 & 3 \\ 5 & 6 & 2 \\ 3 & 4 & -2 \end{pmatrix}.$$

3.3 Application: Solving systems of DEs

Suppose we have a system of differential equations in standard form

$$\begin{cases} x' &= ax + by + f(t), \quad x(0) = x_0, \\ y' &= cx + dy + g(t), \quad y(0) = y_0, \end{cases} \tag{3.7}$$

where a, b, c, d, x_0, y_0 are given constants and $f(t), g(t)$ are given "nice" functions. (Here "nice" will be left vague but basically we don't want these functions to annoy us with any bad behavior while we are trying to solve the differential equations by the method of Laplace transforms.)

One way to solve this system is to take Laplace transforms of both sides. If we let

$$X(s) = \mathcal{L}[x(t)](s), \quad Y(s) = \mathcal{L}[y(t)](s), \quad F(s) = \mathcal{L}[f(t)](s), \quad G(s) = \mathcal{L}[g(t)](s),$$

then (3.7) becomes

$$\begin{cases} sX(s) - x_0 &= aX(s) + bY(s) + F(s), \\ sY(s) - y_0 &= cX(s) + dY(s) + G(s). \end{cases} \tag{3.8}$$

This is now a 2×2 system of linear equations in the unknowns $X(s), Y(s)$ with augmented matrix

$$A = \begin{pmatrix} s-a & -b & F(s)+x_0 \\ -c & s-d & G(s)+y_0 \end{pmatrix}.$$

Compute the row-reduced echelon form of A to solve for $X(s)$ and $Y(s)$. Take inverse Laplace transforms of these to find $x(t)$ and $y(t)$.

Example 3.3.1. Solve

$$\begin{cases} x' &= -y+1, \quad x(0)=0, \\ y' &= -x+t, \quad y(0)=0. \end{cases}$$

The augmented matrix is

$$A = \begin{pmatrix} s & 1 & 1/s \\ 1 & s & 1/s^2 \end{pmatrix}.$$

The row-reduced echelon form of this is

$$\begin{pmatrix} 1 & 0 & 1/s^2 \\ 0 & 1 & 0 \end{pmatrix}.$$

Therefore, $X(s) = 1/s^2$ and $Y(s) = 0$. Taking inverse Laplace transforms, we see that the solution to the system is $x(t) = t$ and $y(t) = 0$. It is easy to check that this is indeed the solution.

To make **Sage** compute the row-reduced echelon form, just type the following:

Sage

```
sage: R = PolynomialRing(QQ,"s")
sage: F = FractionField(R)
sage: s = F.gen()
sage: MS = MatrixSpace(F,2,3)
sage: A = MS([[s,1,1/s],[1,s,1/s^2]])
sage: A.echelon_form()
[     1      0 1/s^2]
[     0      1      0]
```

To make Sage compute the Laplace transform, just type the following:

Sage

```
sage: s,t = var('s,t')
sage: f(t) = 1
sage: laplace(f(t),t,s)
1/s
sage: f(t) = t
sage: laplace(f(t),t,s)
s^(-2)
```

To make Sage compute the inverse Laplace transform, just type the following:

Sage

```
sage: s,t = var('s,t')
sage: F(s) = 1/s^2
sage: inverse_laplace(F(s),s,t)
t
sage: F(s) = 1/(s^2+1)
sage: inverse_laplace(F(s),s,t)
sin(t)
```

Example 3.3.2. The displacement from equilibrium (respectively) for coupled springs attached to a wall on the left (see Figure 3.2)

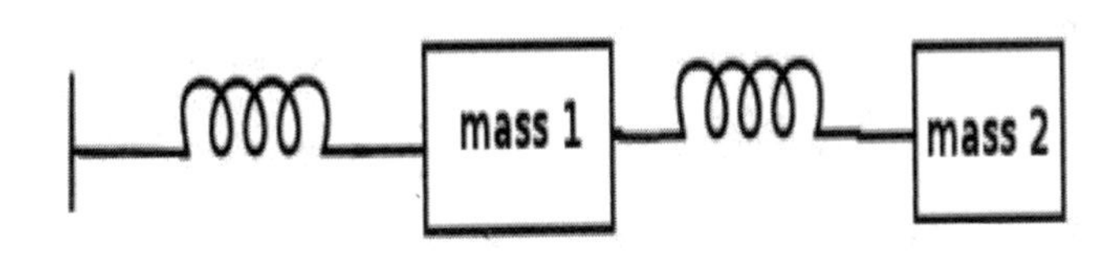

Figure 3.2: A coupled-spring diagram.

is modeled by the system of second-order ODEs

$$m_1x_1'' + (k_1 + k_2)x_1 - k_2x_2 = 0, \quad m_2x_2'' + k_2(x_2 - x_1) = 0,$$

where x_1 denotes the displacement from equilibrium of mass 1, denoted m_1, x_2 denotes the displacement from equilibrium of mass 2, denoted m_2, and k_1, k_2 are the respective spring constants [CS-rref].

As another illustration of solving systems of linear first-order differential equations, we use Sage to solve this problem with $m_1 = 2$, $m_2 = 1$, $k_1 = 4$, $k_2 = 2$, $x_1(0) = 3$, $x_1'(0) = 0$, $x_2(0) = 3$, $x_2'(0) = 0$.

Take Laplace transforms of the first differential equation, $2x_1''(t) + 6x_1(t) - 2x_2(t) = 0$. This says $-2x_1'(0) + 2s^2X_1(s) - 2sx_1(0) - 2X_2(s) + 2X_1(s) = 0$ (where the Laplace transform of a lower-case function is the upper-case function). Take Laplace transforms of the second differential equation, $2x_2''(t) + 2x_2(t) - 2x_1(t) = 0$. This says $s^2X_2(s) + 2X_2(s) - 2X_1(s) - 3s = 0$. Solve these two equations:

Sage

```
sage: s,X,Y = var('s X Y')
sage: eqns = [(2*s^2+6)*X-2*Y == 6*s, -2*X +(s^2+2)*Y == 3*s]
sage: solve(eqns, X,Y)
[[X == (3*s^3 + 9*s)/(s^4 + 5*s^2 + 4),
  Y == (3*s^3 + 15*s)/(s^4 + 5*s^2 + 4)]]
```

This says $X_1(s) = (3s^3 + 9s)/(s^4 + 5s^2 + 4)$, $X_2(s) = (3s^3 + 15s)/(s^4 + 5s^2 + 4)$. Take inverse Laplace transforms to get the answer:

Sage

```
sage: s,t = var('s t')
sage: inverse_laplace((3*s^3 + 9*s)/(s^4 + 5*s^2 + 4),s,t)
cos(2*t) + 2*cos(t)
sage: inverse_laplace((3*s^3 + 15*s)/(s^4 + 5*s^2 + 4),s,t)
4*cos(t) - cos(2*t)
```

Therefore, $x_1(t) = \cos(2t) + 2\cos(t)$, $x_2(t) = 4\cos(t) - \cos(2t)$. Using Sage this can be plotted parametrically using the commands below.

Sage

```
sage: P = parametric_plot([cos(2*t) + 2*cos(t),4*cos(t) - cos(2*t)],0,3)
sage: show(P)
```

The plot is given in Figure 3.3.

Exercises

1. Use the `desolve_system` command in Sage to solve the initial value problem $x' = x + 2y$, $y' = x + 2$, $x(0) = 0$, $y(0) = 0$. Plot the solution.

2. Use the Laplace transform method to solve the initial value problem $x' = x + 2y$, $y' = x + 2$, $x(0) = 0$, $y(0) = 0$.

3. Use the Laplace transform method to solve the initial value problem $x' = x + 2y$, $y' = x + e^t$, $x(0) = 0$, $y(0) = 0$.

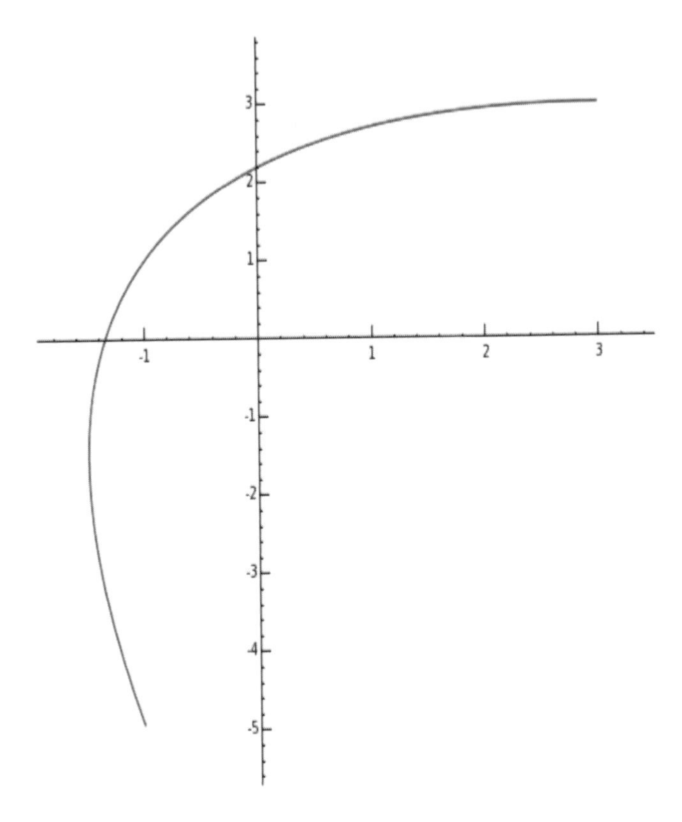

Figure 3.3: Curves $x(t) = \cos(2t) + 2\cos(t)$, $y(t) = 4\cos(t) - \cos(2t)$ along the t-axis.

4. Use the Laplace transform method to solve the initial value problem $x' = x - y$, $y' = y + 1$, $x(0) = 1$, $y(0) = 0$.

3.3.1 Modeling battles using Lanchester's equations

The goal of military analysis is a means of reliably predicting the outcome of military encounters, given some basic information about the forces' status. The case of two combatants in a conventional "directed fire" battle[3] was solved by Frederick William Lanchester,[4] a British engineer who served in the Royal Air Force during World War I. Lanchester discovered a way to model battle-field casualties using systems of differential equations. He

[3]This is a battle where both combatants know roughly where their enemy is and fire in their direction, as in a typical tank battle. This model does not cover the scenario where one side is hidden and firing from cover.

[4]Lanchester was a very remarkable man in many ways, a pioneer British builder of motor boats and cars, and owner of a successful car manufacturing company. Unfortunately, describing his many accomplishments in engineering and operations research woudl take us too far afield in this book.

assumed that if two armies fight, with $x(t)$ troops on one side and $y(t)$ on the other, the rate at which soldiers in one army are put out of action is proportional to the troop strength of their enemy. This give rise to the system of differential equations

$$\begin{cases} x'(t) = -Ay(t), & x(0) = x_0, \\ y'(t) = -Bx(t), & y(0) = y_0, \end{cases}$$

where $A > 0$ and $B > 0$ are constants (called the *fighting effectiveness coefficients*[5]) and x_0 and y_0 are the intial troop strengths. For some historical examples of actual battles modeled using Lanchester's equations, please see references in the excellent paper by McKay [M-intro].

We show here how to solve these equations using Laplace transforms.

Recall the notation for the *hyperbolic cosine* (pronounced "kosh")

$$\cosh(x) = \frac{e^x + e^{-x}}{2},$$

and the *hyperbolic sine* (pronounced "sinch"),

$$\sinh(x) = \frac{e^x - e^{-x}}{2},$$

Example 3.3.3. Solve[6]

$$\begin{cases} x' &= -4y, & x(0) = 400, \\ y' &= -x, & y(0) = 100. \end{cases}$$

This models a battle between "x-men" and "y-men," where the "x-men" die off at a higher rate than the "y-men" (but there are more of them to begin with too).

Take Laplace transforms of both sides of both equations to obtain

$$\begin{cases} sX(s) - 400 = -4Y(s), \\ sY(s) - 100 = -X(s). \end{cases}$$

The augmented matrix of this system (in the unknowns $X(s), Y(s)$) is

$$A = \begin{pmatrix} s & 4 & 400 \\ 1 & s & 100 \end{pmatrix}.$$

The row-reduced echelon form of this is

$$\begin{pmatrix} 1 & 0 & \frac{400(s-1)}{s^2-4} \\ 0 & 1 & \frac{100(s-4)}{s^2-4} \end{pmatrix}.$$

Therefore,

---

[5] U.S. Marine Corp ground troops call this the *kill-shot ratio*.

[6] See also Example 3.4.5.

$$X(s) = 400\frac{s}{s^2-4} - 200\frac{2}{s^2-4}, \quad Y(s) = 100\frac{s}{s^2-4} - 200\frac{2}{s^2-4}.$$

Taking inverse Laplace transforms, we see that the solution to the system is

$$x(t) = 400\cosh(2t) - 200\sinh(2t) = 100e^{2t} + 300e^{-2t}$$

and

$$y(t) = 100\cosh(2t) - 200\sinh(2t) = -50e^{2t} + 150e^{-2t}.$$

The "x-men" win and, in fact,

$$x(0.275) = 346.4102\ldots, \quad y(0.275) = -0.1201\ldots .$$

Question What is $x(t)^2 - 4y(t)^2$? (Hint: It's a constant. Can you explain this?)

To make Sage plot this just type the following:

Sage

```
sage: f = lambda x: 400*cosh(2*x)-200*sinh(2*x)
sage: g = lambda x: 100*cosh(2*x)-200*sinh(2*x)
sage: P = plot(f,0,1)
sage: Q = plot(g,0,1)
sage: show(P+Q)
sage: g(0.275)
 -0.12017933629675781
sage: f(0.275)
 346.41024490088557
```

The graph is given in Figure 3.4.

Here is a similar battle but with different initial conditions.

Example 3.3.4. A battle is modeled by

$$\begin{cases} x' = -4y, & x(0) = 150, \\ y' = -x, & y(0) = 90. \end{cases}$$

(a) Write the solutions in parameteric form. (b) Who wins? When? State the losses for each side.

Solution Take Laplace transforms of both sides:

$$sL[x(t)](s) - x(0) = -4\,L[y(t)](s),$$
$$sL[x(t)](s) - x(0) = -4\,L[y(t)](s).$$

Solving these equations gives

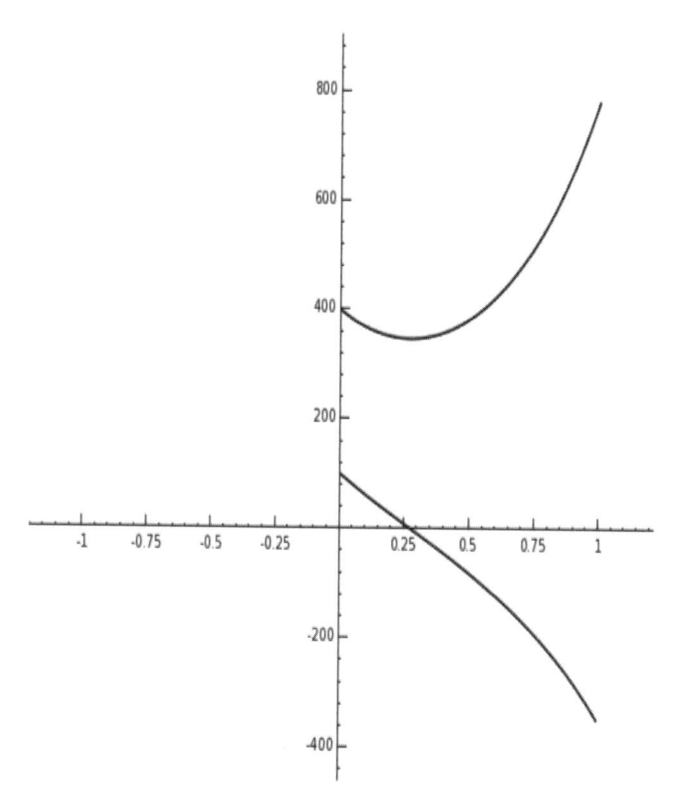

Figure 3.4: Plot of curves $x(t) = 400\cosh(2t) - 200\sinh(2t)$, $y(t) = 100\cosh(2t) - 200\sinh(2t)$ along the t-axis.

$$L[x(t)](s) = \frac{sx(0) - 4y(0)}{s^2 - 4} = \frac{150\,s - 360}{s^2 - 4},$$
$$L[y(t)](s) = -\frac{-sy(0) + x(0)}{s^2 - 4} = -\frac{-90\,s + 150}{s^2 - 4}.$$

Inverting using Laplace transform tables gives

$$x(t) = 150\cosh(2t) - 180\sinh(2t) = -15\,e^{2\,t} + 165\,e^{-2\,t},$$
$$y(t) = 90\,\cosh(2\,t) - 75\,\sinh(2\,t) = \frac{15}{2}e^{2t} + \frac{165}{2}e^{-2t}.$$

Their graph is given in Figure 3.5.

The "y-army" wins. Solving for $x(t) = 0$ gives $t_{win} = \log(11)/4 = .5994738182\ldots$, so the number of survivors is $y(t_{win}) = 49.7493718$; that is, 49 survive.

Lanchester's square law. Suppose that you are more interested in y as a function of x, instead of x and y as functions of t. One can use the chain rule from calculus to derive from

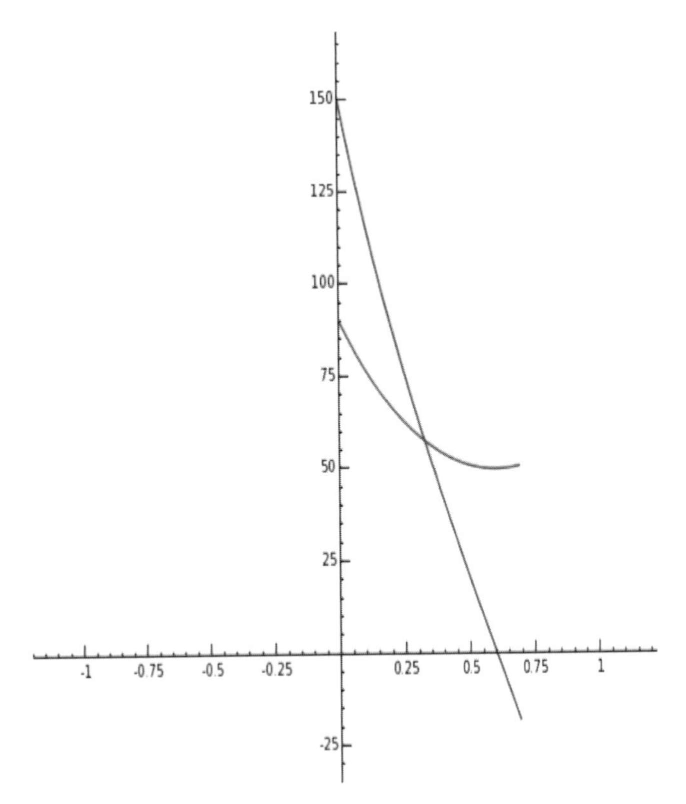

Figure 3.5: Lanchester's model for the x versus y battle.

the system $x'(t) = -Ay(t), y'(t) = -Bx(t)$ the single equation

$$\frac{dy}{dx} = \frac{dy/dt}{dx/dt} = \frac{B}{A}\frac{x}{y}.$$

This differential equation can be solved by the following "separation of variables trick" (due to Lanchester): $Aydy = Bxdx$, so

$$Ay^2 = Bx^2 + C,$$

where C is an unknown constant. (To solve for C you must be given some initial conditions.) The quantity Bx^2 is called the *fighting strength of the x-men* and the quantity Ay^2 is called the *fighting strength of the y-men* ("fighting strength" is not to be confused with "troop strength"). This relationship between the troop strengths is sometimes called *Lanchester's square law* and is sometimes expressed as saying the relative fight strength is a constant:

$$Ay^2 - Bx^2 = \text{constant}.$$

Remark 3.3.1. Here are some comments on those units in support capacity, from [M-intro]. Suppose your total number of troops is some number T, where $x(0)$ are initially in a fighting capacity and $T - x(0)$ are in a support role. If your troops outnumber the enemy then you want to choose the number of support units as the smallest number such that the fighting effectiveness is not decreasing (therefore is roughly constant). The remainder should be actively engaged with the enemy in battle.

A battle between three forces gives rise to the differential equations

$$\begin{cases} x'(t) = -A_1 y(t) - A_2 z(t), & x(0) = x_0, \\ y'(t) = -B_1 x(t) - B_2 z(t), & y(0) = y_0, \\ z'(t) = -C_1 x(t) - C_2 y(t), & z(0) = z_0, \end{cases}$$

where $A_i > 0$, $B_i > 0$, and $C_i > 0$ are constants and x_0, y_0, and z_0 are the intial troop strengths.

Example 3.3.5. Consider the battle modeled by

$$\begin{cases} x'(t) = -y(t) - z(t), & x(0) = 100, \\ y'(t) = -2x(t) - 3z(t), & y(0) = 100, \\ z'(t) = -2x(t) - 3y(t), & z(0) = 100. \end{cases}$$

The x-men and z-men are better fighters than the x-men, in the sense that the coefficient of z in the second differential equation (describing their battle with y) is higher than the coefficient of x, and the coefficient of y in the third differential equation is also higher than the coefficient of x. However, as we will see, the worst fighter wins! (The x-men do have the best defensive abilities, in the sense that A_1 and A_2 are small.)

Taking Laplace transforms (LTs), we obtain the system

$$\begin{cases} sX(s) + Y(s) + Z(s) = 100, \\ 2X(s) + sY(s) + 3Z(s) = 100, \\ 2X(s) + 3Y(s) + sZ(s) = 100, \end{cases}$$

which we solve by row reduction using the augmented matrix

$$\begin{pmatrix} s & 1 & 1 & 100 \\ 2 & s & 3 & 100 \\ 2 & 3 & s & 100 \end{pmatrix}.$$

This has row-reduced echelon form

$$\begin{pmatrix} 1 & 0 & 0 & \frac{100s+100}{s^2+3s-4} \\ 0 & 1 & 0 & \frac{100s-200}{s^2+3s-4} \\ 0 & 0 & 1 & \frac{100s-200}{s^2+3s-4} \end{pmatrix}$$

This means $X(s) = \frac{100s+100}{s^2+3s-4}$ and $Y(s) = Z(s) = \frac{100s-200}{s^2+3s-4}$. Taking inverse LTs, we get the solution: $x(t) = 40e^t + 60e^{-4t}$ and $y(t) = z(t) = -20e^t + 120e^{-4t}$. In other words, the worst fighter wins!

In fact, the battle is over at $t = \log(6)/5 = 0.35\ldots$ and at this time, $x(t) = 71.54\ldots$. Therefore, the worst fighters, the x-men, not only won but have lost fewer than 30% of their men!

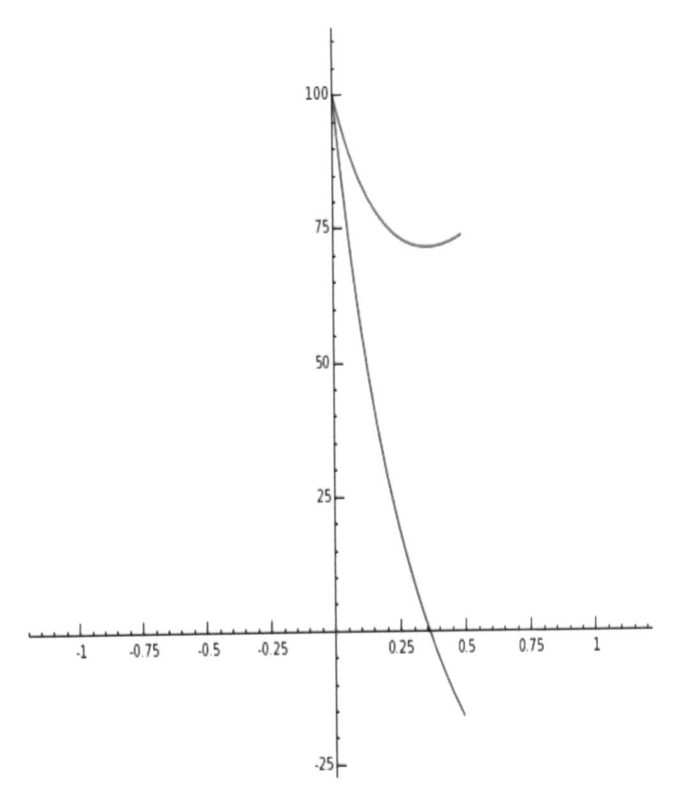

Figure 3.6: Lanchester's model for the x versus y versus z battle.

Exercises

1. Use Sage to solve this: A battle is modeled by

$$\begin{cases} x' = -4y, & x(0) = 150, \\ y' = -x, & y(0) = 40. \end{cases}$$

   (a) Write the solutions in parametric form. (b) Who wins? When? State the losses for each side.

2. All 300 x-men are fighting all 101 y-men, the battle being modeled by

$$\begin{cases} x' = -9y, & x(0) = 300, \\ y' = -x, & y(0) = 101. \end{cases}$$

   Solve this system. Who wins? Compute the losses using Lanchester's square law.

3. All 60 x-men are fighting all 15 y-men, the battle being modeled by

$$\begin{cases} x' = -4y, & x(0) = 60, \\ y' = -x, & y(0) = 15. \end{cases}$$

Solve this system. Who wins? When?

4. Suppose that in the above battle, we don't know the number of y-men:

$$\begin{cases} x' = -4y, & x(0) = 60, \\ y' = -x, & y(0) = y_0. \end{cases}$$

How many y-men are needed to tie? To win?

5. Suppose that in the above battle, we don't know the number of x-men:

$$\begin{cases} x' = -4y, & x(0) = x_0, \\ y' = -x, & y(0) = 15. \end{cases}$$

How many x-men are needed to tie? To win?

6. Suppose that the above battle is modeled instead by

$$\begin{cases} x' = -4xy, & x(0) = 60, \\ y' = -xy, & y(0) = 15. \end{cases}$$

Solve for x in terms of y using the separation of variables trick.

7. Suppose that the above battle is modeled instead by

$$\begin{cases} x' = -4y, & x(0) = 60, \\ y' = -xy, & y(0) = 15. \end{cases}$$

Solve for x in terms of y using the separation of variables trick.

8. Suppose that a battle between three forces is modeled by

$$\begin{cases} x' = -y - 2z, & x(0) = 100, \\ y' = -2x - z, & y(0) = 100, \\ z' = -x - 2y, & y(0) = 100. \end{cases}$$

In other words, the z-men fight better than the x-men, the x-men fight better than the y-men, the y-men fight better than the z-men. Solve for x, y, z.

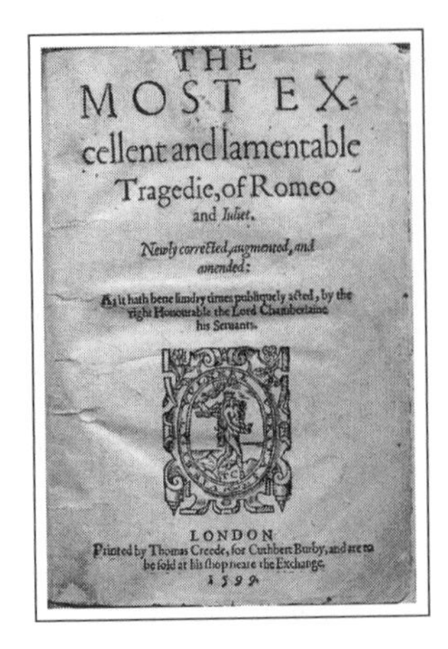

THE
MOST EX-
cellent and lamentable
Tragedie, of Romeo
and Juliet.
Newly corrected, augmented, and
amended:
As it hath bene sundry times publiquely acted, by the
right Honourable the Lord Chamberlaine
his Seruants.
LONDON
Printed by Thomas Creede, for Cuthbert Burby, and are to
be sold at his shop neare the Exchange.
1599.

Figure 3.7: A scene from Shakespeare's *Romeo and Juliet*

3.3.2 Romeo and Juliet

From war and death, we turn to love and romance.

In Figure 3.7, we see the star-crossed lovers[7]

> Romeo:
> If I profane with my unworthiest hand
> This holy shrine, the gentle sin is this:
> My lips, two blushing pilgrims, ready stand
> To smooth that rough touch with a tender kiss.
> Juliet:
> Good pilgrim, you do wrong your hand too much,
> Which mannerly devotion shows in this;
> For saints have hands that pilgrims' hands do touch,
> And palm to palm is holy palmers' kiss.
>
> —*Romeo and Juliet*, Act I, Scene V

William Shakespeare's play *Romeo and Juliet* about two young "star-cross'd lovers" was one of his most popular. The characters Romeo and Juliet represent young lovers whose relationship is, sadly, doomed.

Let $r = r(t)$ denote the love Romeo has for Juliet at time t and let $j = j(t)$ denote the love Juliet has for Romeo at time t.

$$\begin{cases} r' = Aj, & r(0) = r_0, \\ j' = -Br + Cj, & j(0) = j_0, \end{cases} \tag{3.9}$$

[7]Title: *Romeo and Juliet*
Artist: Frank Dicksee
Date: 1884

where $A > 0$, $B > 0$, $C > 0$, r_0, j_0 are given constants. This indicates that Romeo is madly in love with Juliet. The more she loves him, the more he loves her. Juliet is a more complex character. She has eyes for Romeo and her love for him makes her feel good about herself, which makes her love him even more. However, if she senses Romeo seems to love her too much, she reacts negatively and her love wanes.

A few examples illustrate the cyclical nature of the relationship that can arise.

Example 3.3.6. Solve

$$\begin{cases} r' = 5j, & r(0) = 4, \\ j' = -r + 2j, & j(0) = 6. \end{cases}$$

Sage

```
sage: t = var("t")
sage: r = function("r",t)
sage: j = function("j",t)
sage: de1 = diff(r,t) == 5*j
sage: de2 = diff(j,t) == -r+2*j
sage: soln = desolve_system([de1, de2], [r,j],ics=[0,4,6])
sage: rt = soln[0].rhs(); rt
(13*sin(2*t) + 4*cos(2*t))*e^t
sage: jt = soln[1].rhs(); jt
(sin(2*t) + 6*cos(2*t))*e^t
```

To solve this using Laplace transforms, take Laplace transforms of both sides, to obtain

$$sR(s) - 4 = 5J(s), \qquad sJ(s) - 6 = -R(s) + 2J(s),$$

where $R(s) = \mathcal{L}[r(t)](s)$, $J(s) = \mathcal{L}[j(t)](s)$. This gives rise to the augmented matrix

$$\left(\begin{array}{ccc} s & -5 & 4 \\ 1 & s-2 & 6 \end{array} \right).$$

Computing the row-reduced echelon form gives

$$R(s) = -10\frac{2-3s}{s(s^2-2s+5)} + \frac{4}{s}, \quad J(s) = -2\frac{2-3s}{(s^2-2s+5}.$$

Taking inverse Laplace transforms gives

$$r(t) = (13\sin(2t) + 4\cos(2t))e^t, \qquad j(t) = (\sin(2t) + 6\cos(2t))e^t.$$

The parametric plot of $x = r(t)$, $y = j(t)$ is given in Figure 3.8.

Example 3.3.7. Solve

$$\begin{cases} r' = 10j, & r(0) = 4, \\ j' = -r + j/10, & j(0) = 6. \end{cases}$$

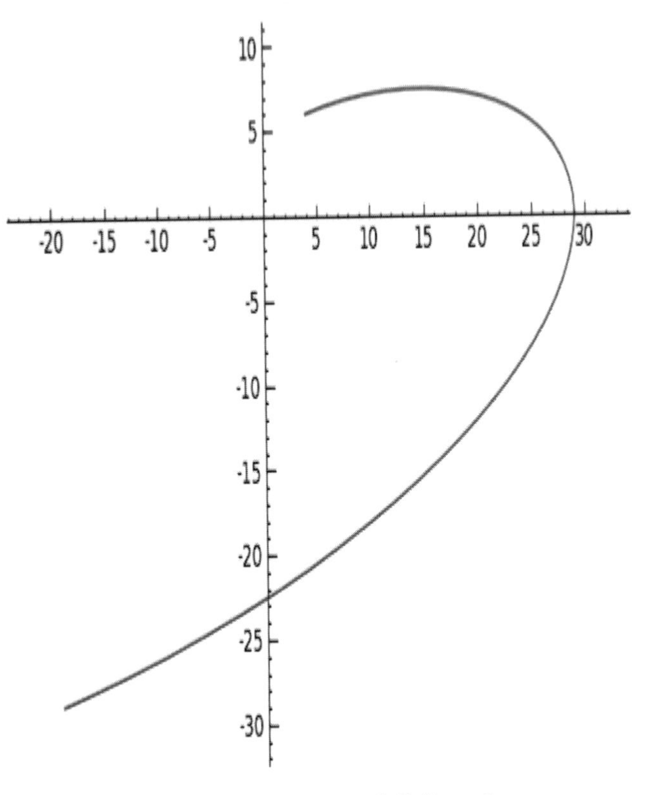

Figure 3.8: Romeo and Juliet plots.

To solve this using Laplace transforms, take Laplace transforms of both sides, to obtain

$$sR(s) - 4 = 10J(s), \qquad sJ(s) - 6 = -R(s) + \frac{1}{10}J(s),$$

where $R(s) = \mathcal{L}[r(t)](s)$, $J(s) = \mathcal{L}[j(t)](s)$. Solving this as above gives

$$r(t) = \frac{4}{3999}\left(299\sqrt{3999}\sin\left(\frac{1}{20}\sqrt{3999}t\right) + 3999\cos\left(\frac{1}{20}\sqrt{3999}t\right)\right)e^{t/20},$$

$$j(t) = -\frac{2}{3999}\left(37\sqrt{3999}\sin\left(\frac{1}{20}\sqrt{3999}t\right) - 11997\cos\left(\frac{1}{20}\sqrt{3999}t\right)\right)e^{t/20}$$

Sage

```
sage: t = var("t")
sage: r = function("r",t)
sage: j = function("j",t)
sage: de1 = diff(r,t) == 10*j
sage: de2 = diff(j,t) == -r+(1/10)*j
sage: soln = desolve_system([de1, de2], [r,j],ics=[0,4,6])
sage: rt = soln[0].rhs(); rt
```

```
4/3999*(299*sqrt(3999)*sin(1/20*sqrt(3999)*t) + 3999*cos(1/20*sqrt(3999)*t))*e^(1/20*t)
sage: jt = soln[1].rhs(); jt
-2/3999*(37*sqrt(3999)*sin(1/20*sqrt(3999)*t) - 11997*cos(1/20*sqrt(3999)*t))*e^(1/20*t)
```

The parametric plot of $x = r(t)$, $y = j(t)$ is given in Figure 3.9.

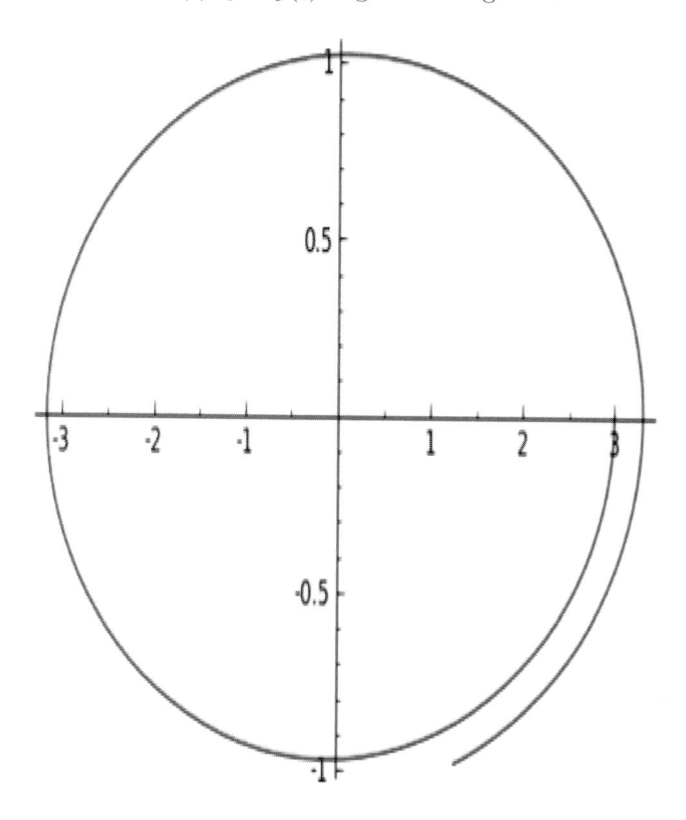

Figure 3.9: A Romeo and Juliet plot.

Exercises

1. Rewrite the second-order differential equation $x'' + 3x' + 5x = t$ as a system of first-order differential equations. (You do not have to find the solution.)

2. Find the general solution to the system $x_1' = x_1 + 2x_2$, $x_2' = 3x_1 + 2x_2$.

3. Find the general solution to the system $x_1' = x_1 - 5x_2$, $x_2' = x_1 - x_2$. Sketch some of the solutions near the origin.

4. Solve the initial value problem $x_1' = x_1 + 2x_2$, $x_2' = -2x_1 + x_2$, $x_1(0) = 1$, $x_2(0) = 0$.

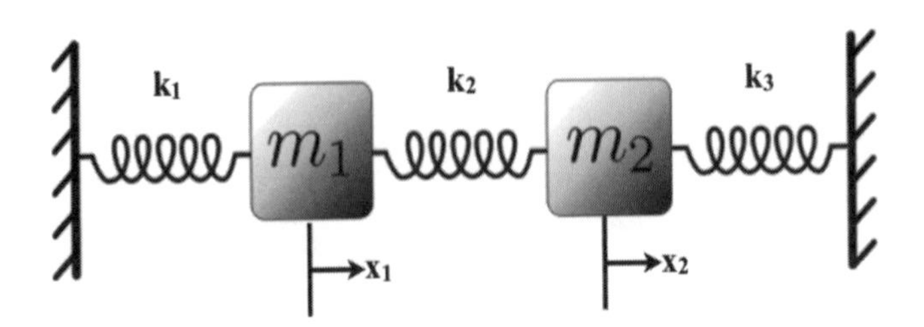

Figure 3.10: Coupled-spring diagram.

For the next two problems, consider two blocks of mass m_1 and m_2 connected by springs to each other and to walls as shown in Figure 3.10. The displacements of the masses from their equilibrium positions are denoted by x_1 and x_2. The stiffnesses of the three springs are k_1, k_2, and k_3 as shown. Compute the natural frequencies and describe the natural modes of oscillation in each of the two following cases:

5. $k_1 = k_3 = 0$ and $k_2 = 4$, and $m_1 = m_2 = 1$.
6. $k_1 = k_3 = 1$ and $k_2 = 4$, and $m_1 = m_2 = 1$.
7. Suppose a rocket is fired and after its initial burn it is at an altitude of $h = 5$ km with a velocity of 4 km/s straight up. Since it may reach a significant altitude, it could be too inaccurate to use a constant gravitational force. Instead, we will neglect air resistance but use the Newtonian law of gravity:

$$\frac{d^2h}{dt^2} = -\frac{gR^2}{(h+R)^2},$$

where h is the height above sea level in kilometers, $R = 6378$ km, and $g = 0.0098$ km/s^2.

Convert this to a system of first-order equations and use a numerical method to find the maximum height of the rocket. Your answer should have two digits of accuracy (within the assumptions of the model). Use of `Sage` is highly encouraged for this problem.

8. Use `Sage` to analyze the problem

$$\begin{cases} r' = 53j/10, & r(0) = 3, \\ j' = -r + j/5, & j(0) = 6. \end{cases}$$

9. (*) Use `Sage` to analyze the Lotka-Volterra/predator-prey model

$$\begin{cases} x' = x(-1+2y), & x(0) = 4, \\ y' = y(-3x+4y), & y(0) = 6. \end{cases}$$

3.3.3 Electrical networks using Laplace transforms

Suppose we have an electrical network (i.e., a series of electrical circuits) involving emfs (electromotive forces or batteries), resistors, capacitors, and inductors. We use the dictionary in Figure 2.13 to translate between the network diagram and the differential equations.

EE object	term in DE (the voltage drop)	units	symbol
charge	$q = \int i(t)\,dt$	coulombs	
current	$i = q'$	amps	
emf	$e = e(t)$	volts V	⊣ ⊢ (+)
resistor	$Rq' = Ri$	ohms Ω	resistor (zigzag)
capacitor	$C^{-1}q$	farads	⊣ ⊢
inductor	$Lq'' = Li'$	henries	coil

Also, recall from §2.8 Kirchoff's laws.

Example 3.3.8. Consider the simple RC circuit given by the diagram in Figure 3.11.

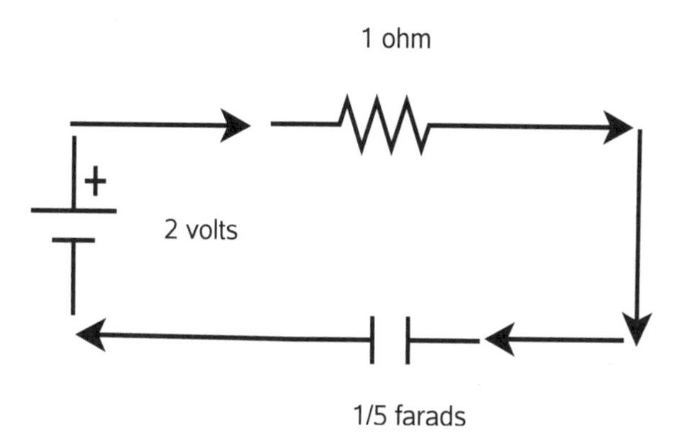

Figure 3.11: A simple circuit.

According to Kirchoff's second law and the dictionary of Figure 2.13, this circuit corresponds to the differential equation

$$q' + 5q = 2.$$

The general solution to this is $q(t) = 1 + ce^{-2t}$, where c is a constant which depends on the initial charge on the capacitor.

Remark 3.3.2. The convention of assuming that electricity flows from positive to negative on the terminals of a battery is referred to as "conventional flow." The physically correct but opposite assumption is referred to as "electron flow." We shall assume the electron flow convention.

Example 3.3.9. Consider the network given by the diagram in Figure 3.12. Assume the initial charges are 0.

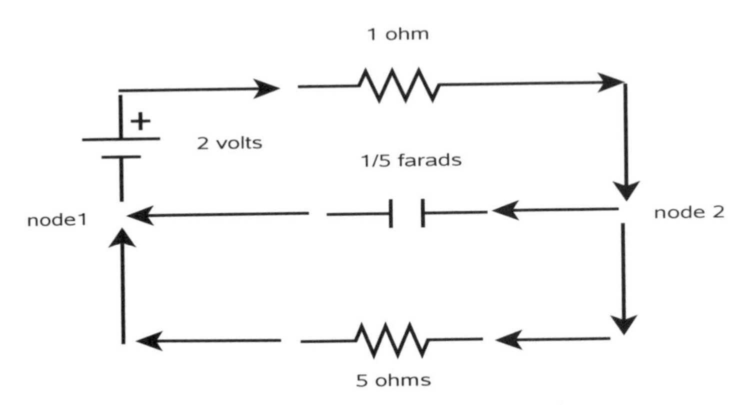

Figure 3.12: A network.

One difference between the circuit in Figure 3.11 and the network in Figure 3.12 is that the charges on the three paths between the two nodes (labeled node 1 and node 2 for convenience) must be labeled. The charge passing through the 5 Ω resistor we label q_1, the charge on the capacitor we denote by q_2, and the charge passing through the 1 Ω resistor we label q_3.

There are three closed loops in the diagram in Figure 3.12: the "top loop," the "bottom loop," and the "big loop." The loops will be traversed in the clockwise direction. Note the top loop looks like the simple circuit given in Example 3.3.8. However, it cannot be solved in the same way, since the current passing through the 5 ohm resistor will affect the charge on the capacitor. This current is not present in the circuit of Example 3.3.8 but it does occur in the network above.

Kirchoff's laws and the dictionary of Figure 2.13 give

$$\begin{cases} q_3' + 5q_2 = 2, & q_1(0) = 0, \\ 5q_1' - 5q_2 = 0, & q_2(0) = 0, \\ 5q_1' + q_3' = 2, & q_3(0) = 0. \end{cases} \tag{3.10}$$

A few things to take note of:

- The minus sign in front of the term associated with the capacitor ($-5q_2$). This is because we are going clockwise, against the direction of the current.

- The equation associated with the big loop (the last equation displayed above) is the sum of the first two equations. Also, the big loop is roughly speaking the sum of the top and the bottom loops.

Kirchoff's first law says that the sum of the currents into a node is zero. At node 1, this says $q_1' + q_2' - q_3' = 0$, or $q_3' = q_1' + q_2'$. Since $q_1(0) = q_2(0) = q_3(0) = 0$, this implies $q_3 = q_1 + q_2$. After taking Laplace transforms of the three differential equations above, we get

$$sQ_3(s) + 5Q_2(s) = 2/s, \qquad 5sQ_1(s) - 5Q_2(s) = 0.$$

Note that you don't need to take the Laplace transform of the second equation since it is the difference of the other equations. The Laplace transform of $q_1 + q_2 = q_3$ (Kirchoff's law) gives $Q_1(s) + Q_2(s) - Q_3(s) = 0$. Therefore, taking Laplace transforms of Kirchoff's law and the first and third equations in (3.10) gives this matrix equation

$$\begin{pmatrix} 0 & 5 & s \\ 5s & 0 & s \\ 1 & 1 & -1 \end{pmatrix} \begin{pmatrix} Q_1(s) \\ Q_2(s) \\ Q_3(s) \end{pmatrix} = \begin{pmatrix} 2/s \\ 2/s \\ 0 \end{pmatrix}.$$

The augmented matrix describing this system is

$$\begin{pmatrix} 0 & 5 & s & 2/s \\ 5s & 0 & s & 2/s \\ 1 & 1 & -1 & 0 \end{pmatrix}.$$

The row-reduced echelon form is

$$\begin{pmatrix} 1 & 0 & 0 & 2/(s^3+6s^2) \\ 0 & 1 & 0 & 2/(s^2+6s) \\ 0 & 0 & 1 & 2(s+1)/(s^3+6s^2) \end{pmatrix}.$$

Therefore

$$Q_1(s) = \frac{2}{s^3+6s^2}, \qquad Q_2(s) = \frac{2}{s^2+6s}, \qquad Q_3(s) = \frac{2(s+1)}{s^2(s+6)}.$$

This implies

$$q_1(t) = -1/18 + e^{-6t}/18 + t/3, \qquad q_2(t) = 1/3 - e^{-6t}/3, \qquad q_3(t) = q_2(t) + q_1(t).$$

This computation can be done in Sage as well:

Sage

```
sage: s = var("s")
sage: MS = MatrixSpace(SymbolicExpressionRing(), 3, 4)
sage: A = MS([[0,5,s,2/s],[5*s,0,s,2/s],[1,1,-1,0]])
sage: B = A.echelon_form(); B

[   1      0      0    2/(5*s^2) - (-2/(5*s) - 2/(5*s^2))/(5*(-s/5 - 6/5))]
[   0      1      0   2/(5*s) - (-2/(5*s) - 2/(5*s^2))*s/(5*(-s/5 - 6/5)) ]
[   0      0      1              (-2/(5*s) - 2/(5*s^2))/(-s/5 - 6/5)      ]

sage: Q1 = B[0,3]
```

```
sage: t = var("t")
sage: Q1.inverse_laplace(s,t)
e^(-(6*t))/18 + t/3 - 1/18
sage: Q2 = B[1,3]
sage: Q2.inverse_laplace(s,t)
1/3 - e^(-(6*t))/3
sage: Q3 = B[2,3]
sage: Q3.inverse_laplace(s,t)
-5*e^(-(6*t))/18 + t/3 + 5/18
```

Example 3.3.10. Consider the network given by the diagram in Figure 3.13. Assume the initial currents are 0.

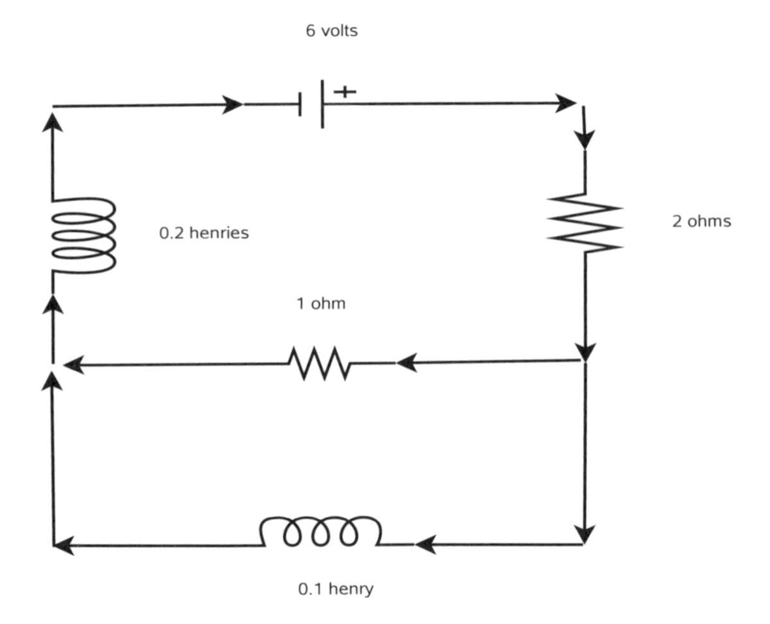

Figure 3.13: Another network.

Using Kirchoff's laws, you get a system

$$\begin{cases} i_1 - i_2 - i_3 = 0, \\ 2i_1 + i_2 + (0.2)i_1' = 6, \\ (0.1)i_3' - i_2 = 0. \end{cases}$$

Take Laplace transforms of each of these three differential equations. You get a 3×3 system in the unknowns $I_1(s) = \mathcal{L}[i_1(t)](s)$, $I_2(s) = \mathcal{L}[i_2(t)](s)$, and $I_3(s) = \mathcal{L}[i_3(t)](s)$. The augmented matrix of this system is

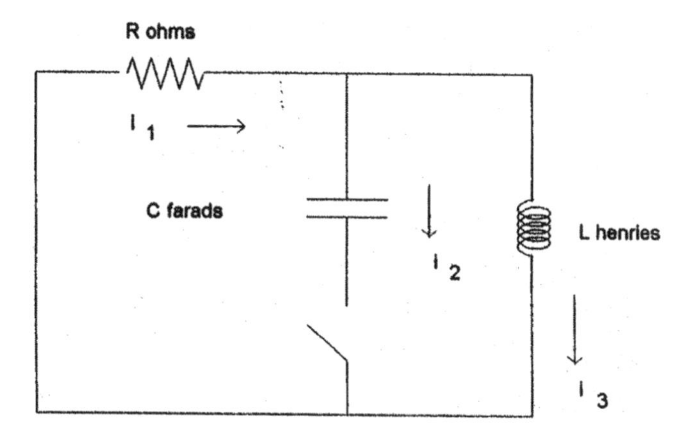

Figure 3.14: Network diagram.

$$\begin{pmatrix} 1 & -1 & -1 & 0 \\ 2+s/5 & 1 & 0 & 6/s \\ 0 & -1 & s/10 & 0 \end{pmatrix}.$$

(Check this yourself!) The row-reduced echelon form is

$$\begin{pmatrix} 1 & 0 & 0 & \frac{30(s+10)}{s(s^2+25s+100)} \\ 0 & 1 & 0 & \frac{30}{s^2+25s+100} \\ 0 & 0 & 1 & \frac{300}{s(s^2+25s+100)} \end{pmatrix}.$$

Therefore

$$I_1(s) = -\frac{1}{s+20} - \frac{2}{s+5} + \frac{3}{s}, \quad I_2(s) = -\frac{3}{s+20} + \frac{3}{s+5}, \quad I_3(s) = \frac{1}{s+20} - \frac{4}{s+5} + \frac{3}{s}.$$

This implies

$$i_1(t) = 3 - 2e^{-5t} - e^{-20t}, \quad i_2(t) = 3e^{-5t} - 3e^{-20t}, \quad i_3(t) = 3 - 4e^{-5t} + e^{-20t}.$$

Exercises

1. Use Sage to solve for $i_1(t)$, $i_2(t)$, and $i_3(t)$ in the network diagram given in Figure 3.14.

2. (a) For the circuit in Figure 3.14 show that the charge q on the capacitor and the current i_3 in the right branch satisfy the system of differential equations

$$q' + (1/RC)q + i_3 = 0,$$

$$i_3' + (1/LC)q = 0.$$

(b) When the switch in the circuit is closed at time $t = 0$, the current i_3 is 0 A and the charge on the capacitor is 5 C. With $R = 2$, $L = 3$, $C = 1/6$ use Laplace transforms to find the charge $q(t)$ on the capacitor.

3.4 Eigenvalue method for systems of DEs

3.4.1 Motivation

First, we shall try to motivate the study of eigenvalues and eigenvectors. This section hopefully will convince you that they are very natural.

Let A be any $n \times n$ matrix. If v is a vector in $\mathbb{R}^n$ then Av is another vector in $\mathbb{R}^n$. more generally, we see that the "multiplication by A map,"

$$A : \mathbb{R}^n \to \mathbb{R}^n,$$
$$v \mapsto Av,$$

defines a function. This function is by no means arbrtrary. It is a rather special type of function, with a lot of useful properties. The only property that is important now is that there are n complex numbers, and (usually) n linearly independent complex vectors, which together determine a lot of the important properties of A. These numbers are the eigenvalues of A and these vectors are the eigenvectors of A.

The simplest matrices A for which the eigenvalues and eigenvectors are easy to compute are the diagonal matrices.

The arithmetic of diagonal matrices.

We'll focus for simplicity on the 2×2 case, but everything applies to the general case of $n \times n$ diagonal matrices.

- Addition is easy:

$$\begin{pmatrix} a_1 & 0 \\ 0 & a_2 \end{pmatrix} + \begin{pmatrix} b_1 & 0 \\ 0 & b_2 \end{pmatrix} = \begin{pmatrix} a_1 + b_1 & 0 \\ 0 & a_2 + b_2 \end{pmatrix}.$$

- Multiplication is easy:

$$\begin{pmatrix} a_1 & 0 \\ 0 & a_2 \end{pmatrix} \cdot \begin{pmatrix} b_1 & 0 \\ 0 & b_2 \end{pmatrix} = \begin{pmatrix} a_1 \cdot b_1 & 0 \\ 0 & a_2 \cdot b_2 \end{pmatrix}.$$

- Powers are easy:

$$\begin{pmatrix} a_1 & 0 \\ 0 & a_2 \end{pmatrix}^n = \begin{pmatrix} a_1^n & 0 \\ 0 & a_2^n \end{pmatrix}.$$

- You can even exponentiate diagonal matrices using the usual power series expansion for $exp(x) = e^x = 1 + x + x^2/2 + \ldots$:

$$\begin{aligned}
\exp\left(t\begin{pmatrix} a_1 & 0 \\ 0 & a_2 \end{pmatrix}\right) &= \begin{pmatrix} 1 & 0 \\ 0 & 1 \end{pmatrix} + t\begin{pmatrix} a_1 & 0 \\ 0 & a_2 \end{pmatrix} \\
&\quad + \frac{1}{2!}t^2\begin{pmatrix} a_1 & 0 \\ 0 & a_2 \end{pmatrix}^2 + \frac{1}{3!}t^3\begin{pmatrix} a_1 & 0 \\ 0 & a_2 \end{pmatrix}^3 + \ldots \\
&= \begin{pmatrix} 1 & 0 \\ 0 & 1 \end{pmatrix} + \begin{pmatrix} ta_1 & 0 \\ 0 & ta_2 \end{pmatrix} \\
&\quad + \begin{pmatrix} \frac{1}{2!}t^2a_1^2 & 0 \\ 0 & \frac{1}{2!}t^2a_2^2 \end{pmatrix} + \begin{pmatrix} \frac{1}{3!}t^3a_1^3 & 0 \\ 0 & \frac{1}{3!}t^3a_2^3 \end{pmatrix} + \ldots \\
&= \begin{pmatrix} e^{ta_1} & 0 \\ 0 & e^{ta_2} \end{pmatrix}.
\end{aligned}$$

So, it is relatively easy to compute with diagonal matrices.

The eigenvalues of a diagonal matrix A are the numbers on the diagonal of A (namely, a_1 and a_2 in the example above). The eigenvectors of a diagonal matrix A are (provided the diagonal entries are all nonzero) the standard basis vectors.

Geometric interpretation of matrix conjugation.

You and your friend are piloting a rocket in space. You handle the controls, your friend handles the map. To communicate, you have to "change coordinates." Your coordinates are those of the rocketship (straight ahead is one direction, to the right is another). Your friend's coordinates are those of the map (north and east are map directions). Changing coordinates corresponds algebraically to conjugating by a suitable matrix. Using an example, we'll see how this arises in a specific case.

Your basis vectors are

$$v_1 = (1,0), \quad v_2 = (0,1),$$

which we call the "v-space coordinates," and the map's basis vectors are

$$w_1 = (1,1), \quad w_2 = (1,-1),$$

which we call the "w-space coordinates."

For example, the point $(7,3)$ is, in v-space coordinates of course $(7,3)$ but in the w-space coordinates, $(5,2)$ since $5w_1 + 2w_2 = 7v_1 + 3v_2$. Indeed, the matrix $A = \begin{pmatrix} 1 & 1 \\ 1 & -1 \end{pmatrix}$ sends $\begin{pmatrix} 5 \\ 2 \end{pmatrix}$ to $\begin{pmatrix} 7 \\ 3 \end{pmatrix}$.

Suppose we flip about the 45° line (the "diagonal") in each coordinate system. In the v-space:

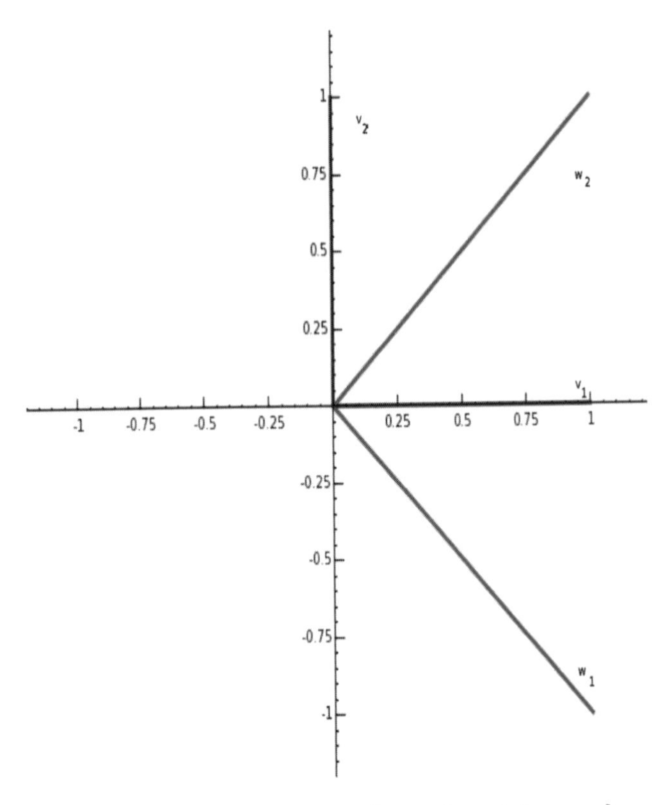

Figure 3.15: Basis vectors $\{v_1, v_2\}$ and $\{w_1, w_2\}$.

$$av_1 + bv_2 \longmapsto bv_1 + av_2,$$
$$\begin{pmatrix} a \\ b \end{pmatrix} \longmapsto \begin{pmatrix} 0 & 1 \\ 1 & 0 \end{pmatrix} \begin{pmatrix} a \\ b \end{pmatrix}.$$

In other words, in v-space, the flip map is $\begin{pmatrix} 0 & 1 \\ 1 & 0 \end{pmatrix}$.

In the w-space:

$$wv_1 + wv_2 \longmapsto aw_1 - bw_2,$$
$$\begin{pmatrix} a \\ b \end{pmatrix} \longmapsto \begin{pmatrix} 1 & 0 \\ 0 & -1 \end{pmatrix} \begin{pmatrix} a \\ b \end{pmatrix}.$$

In other words, in w-space, the flip map is $\begin{pmatrix} 1 & 0 \\ 0 & -1 \end{pmatrix}$.

Conjugating by the matrix A converts the flip map in w-space to the flip map in v-space:

$$A \cdot \begin{pmatrix} 1 & 0 \\ 0 & -1 \end{pmatrix} \cdot A^{-1} = \begin{pmatrix} 0 & 1 \\ 1 & 0 \end{pmatrix}.$$

3.4.2 Computing eigenvalues

At first glance, matrices can seem to be complicated objects. When the matrix is a diagonal matrix it is relatively simple. In the case of diagonal matrices, the eigenvalues are the diagonal entries. It turns out that "most" matrices are conjugate to a diagonal matrix.

The eigenvalues of a general square matrix

- are relatively easy to compute, and
- tell us something about the "behavior" of the matrix.

This is a bit vague but more details will be given. For us, as you will see, eigenvalues are very useful for solving systems of linear first-order differential equations.

Each $n \times n$ matrix A has exactly n (counted according to multiplicity) eigenvalues λ_1, λ_2, ..., λ_n. These numbers are the roots of the *characteristic polynomial*

$$p(\lambda) = p_A(\lambda) = \det(A - \lambda I). \tag{3.11}$$

In other words, subtract the variable λ from each diagonal entry of A and take the determinant. This is a polynomial of degree n in λ. A corresponding *eigenvector* is any nonzero vector $\vec{v}$ which satisfies the (defining) *eigenvector equation*

$$A\vec{v} = \lambda\vec{v}, \tag{3.12}$$

where λ is a fixed eigenvalue. Note that you can "scale" an eigenvector $\vec{v}$ (i.e., replace $\vec{v}$ by $c\vec{v}$, for some scalar $c \neq 0$) and get another eigenvector.

Example 3.4.1. Consider an $n \times n$ diagonal matrix. The standard basis elements ($e_1 = (1, 0, \dots, 0)$, ..., $e_n = (0, \dots, 0, 1)$) are the eigenvectors and the diagonal elements are the eigenvalues.

Example 3.4.2. Find the eigenvalues and eigenvectors of

$$A = \begin{pmatrix} 0 & -1 & 1 \\ -4 & 0 & 2 \\ 0 & 0 & 3 \end{pmatrix}.$$

We compute

$$p(\lambda) = \det \begin{pmatrix} 0-\lambda & -1 & 1 \\ -4 & -\lambda & 2 \\ 0 & 0 & 3-\lambda \end{pmatrix} = -(\lambda - 2)(\lambda + 2)(\lambda - 3).$$

Therefore, $\lambda_1 = -2$, $\lambda_2 = 2$, $\lambda_3 = 3$. We solve for the corresponding eigenvectors, $\vec{v}_1$, $\vec{v}_2$, $\vec{v}_3$. For example, let

$$\vec{v}_3 = \begin{pmatrix} x \\ y \\ z \end{pmatrix}.$$

We solve for x, y, z in the eigenvector equation

$$\begin{pmatrix} 0-\lambda & -1 & 1 \\ -4 & -\lambda & 2 \\ 0 & 0 & 3-\lambda \end{pmatrix} \begin{pmatrix} x \\ y \\ z \end{pmatrix} = 3 \begin{pmatrix} x \\ y \\ z \end{pmatrix}.$$

This gives rise to the system of linear equations

$$\begin{array}{rrrl} & -y & +z & = 3x, \\ -4x & & +2z & = 3y, \\ & & 3z & = 3z. \end{array}$$

You can find some nonzero solution to these using row reduction/Gauss elimination, for example. The row-reduced echelon form of

$$\begin{pmatrix} -3 & -1 & 1 & 0 \\ -4 & -3 & 2 & 0 \\ 0 & 0 & 0 & 0 \end{pmatrix}$$

is

$$A = \begin{pmatrix} 1 & 0 & -1/5 & 0 \\ 0 & 1 & -2/5 & 0 \\ 0 & 0 & 0 & 0 \end{pmatrix},$$

so z is anything (nonzero, that is), $y = 2z/5$, and $x = z/5$. One nonzero solution is $x = 1$, $y = 2$, $z = 5$, so

$$\vec{v}_3 = \begin{pmatrix} 1 \\ 2 \\ 5 \end{pmatrix}.$$

However, any other scalar multiple of this will also satisfy the eigenvector equation. The other eigenvectors can be computed similarly. We obtain in this way

$$\vec{v}_1 = \begin{pmatrix} 1 \\ 2 \\ 0 \end{pmatrix}, \qquad \vec{v}_2 = \begin{pmatrix} 1 \\ -2 \\ 0 \end{pmatrix}.$$

Since this section is only intended to be motivation, we shall not prove this here (see any text on linear algebra, for example [B-rref] or [H-rref]).

Example 3.4.3. Sage can compute with complex numbers. You can use "exact arithmetic":

Sage

```
sage: MS = MatrixSpace(QQ[I],2,2)
sage: A = MS([[0, 1], [-1, 0]])
sage: A.charpoly()
x^2 + 1
sage: A.eigenspaces_right()
```

```
[
(I, Vector space of degree 2 and dimension 1 over Number Field in I
with defining polynomial x^2 + 1
User basis matrix:
[1 I]),
(-I, Vector space of degree 2 and dimension 1 over Number Field in I
with defining polynomial x^2 + 1
User basis matrix:
[ 1 -I])
]
```

Or you can use "floating-point arithmetic":

Sage

```
sage: MS = MatrixSpace(CC,2,2)
sage: A = MS([[0,1],[-1,0]])
sage: A.eigenspaces_right()

[
(1.00000000000000*I, Vector space of degree 2 and dimension 1
over Complex Field with 53 bits of precision
User basis matrix:
[  1.00000000000000 1.00000000000000*I]),
(-1.00000000000000*I, Vector space of degree 2 and dimension 1
over Complex Field with 53 bits of precision
User basis matrix:
[   1.00000000000000 -1.00000000000000*I])
]
sage: p = A.charpoly(); p
x^2 + 1.00000000000000
sage: p.roots()
[(-1.00000000000000*I, 1), (1.00000000000000*I, 1)]
```

Exercises

Find the eigenvalues and eigenvectors of the following matrices by hand:

1. $\begin{pmatrix} 4 & -2 \\ 1 & 1 \end{pmatrix}$.

2. $\begin{pmatrix} 5 & -6 \\ 3 & -4 \end{pmatrix}$.

3. $\begin{pmatrix} 2 & 0 & 0 \\ 5 & 3 & -2 \\ 2 & 0 & 1 \end{pmatrix}$.

4. $\begin{pmatrix} 3 & 1 & 1 \\ 0 & 1 & 0 \\ 0 & 0 & 1 \end{pmatrix}$.

5. $\begin{pmatrix} 0 & -2 \\ 1 & 0 \end{pmatrix}$.

6. $\begin{pmatrix} 0 & 1 \\ -1 & 0 \end{pmatrix}$.

Use Sage to find eigenvalues and eigenvectors of the following matrices

7. $\begin{pmatrix} 1 & -1 \\ 4 & 1 \end{pmatrix}$.

8. $\begin{pmatrix} -2 & 3 \\ -3 & 4 \end{pmatrix}$.

9. $\begin{pmatrix} 1 & -1 & 0 \\ 4 & 1 & 0 \\ 0 & 0 & -13 \end{pmatrix}$.

10. Show that if A is invertible and λ is an eigenvalue of A, then $1/\lambda$ is an eigenvalue of A^{-1}. Are the eigenvectors the same?

11. If λ is an eigenvalue for a matrix A and μ is an eigenvalue for a matrix B, is $\lambda + \mu$ an eigenvalue for the matrix $A + B$? If not, find a counterexample.

12. Find a matrix P such that $P^{-1}AP = D$, where D is a diagonal matrix, for the following two matrices if such a P exists.

    - $A = \begin{pmatrix} 0 & 1 & 0 \\ -1 & 2 & 0 \\ -1 & 1 & 1 \end{pmatrix}$.
    - $A = \begin{pmatrix} 1 & 0 & 0 & 1 \\ 0 & 1 & 0 & 1 \\ 0 & 0 & 1 & 1 \\ 0 & 0 & 0 & 2 \end{pmatrix}$.

13. By computing the eigenvalues and eigenvectors of $A = \begin{pmatrix} 3 & -2 \\ 1 & 0 \end{pmatrix}$ find a matrix P such that $P^{-1}AP = D$ where D is a diagonal matrix. Use this diagonalization to compute A^5.

3.4.3 The eigenvalue method

We next discuss a method for solving a linear homogenous system of differential equations.

Solution strategy.

Problem. Solve

$$\begin{cases} x' = ax + by, & x(0) = x_0, \\ y' = cx + dy, & y(0) = y_0. \end{cases}$$

Solution. Let

$$A = \begin{pmatrix} a & b \\ c & d \end{pmatrix}.$$

In matrix notation, the system of differential equations becomes

$$\vec{X}' = A\vec{X}, \quad \vec{X}(0) = \begin{pmatrix} x_0 \\ y_0 \end{pmatrix}, \tag{3.13}$$

where $\vec{X} = \vec{X}(t) = \begin{pmatrix} x(t) \\ y(t) \end{pmatrix}$. Recall the manner in which we solved homogeneous constant-coefficient second-order ODEs $ax'' + bx' + cx = 0$. We used "Euler's guess" $x = Ce^{rt}$. Similarly, for (3.13), we try to guess an exponential: $\vec{X}(t) = \vec{c}e^{\lambda t}$. Here, λ is used instead of r to stick with notational convention, and $\vec{c}$ in place of C since we need a constant vector. Plugging this guess into the matrix differential equation $\vec{X}' = A\vec{X}$ gives $\lambda\vec{c}e^{\lambda t} = A\vec{c}e^{\lambda t}$, or (cancelling $e^{\lambda t}$)

$$A\vec{c} = \lambda\vec{c}.$$

This means that λ is an eigenvalue of A with eigenvector $\vec{c}$.

- Find the eigenvalues. These are the roots of the characteristic polynomial

  $$p(\lambda) = \det \begin{pmatrix} a - \lambda & b \\ c & d - \lambda \end{pmatrix} = \lambda^2 - (a + d)\lambda + (ad - bc).$$

  Call them λ_1, λ_2 (in any order you like).

  You can use the quadratic formula, for example, to get them:

  $$\lambda_1 = \frac{a+d}{2} + \frac{\sqrt{(a+d)^2 - 4(ad-bc)}}{2}, \quad \lambda_2 = \frac{a+d}{2} - \frac{\sqrt{(a+d)^2 - 4(ad-bc)}}{2}.$$

- Find the eigenvectors. If $b \neq 0$ then you can use the formulas

  $$\vec{v}_1 = \begin{pmatrix} b \\ \lambda_1 - a \end{pmatrix}, \qquad \vec{v}_2 = \begin{pmatrix} b \\ \lambda_2 - a \end{pmatrix}. \tag{3.14}$$

  In general, you can get them by solving the *eigenvector equation* $A\vec{v} = \lambda\vec{v}$.

- Plug these into the following formulas:

(a) $\lambda_1 \neq \lambda_2$, real:

$$\begin{pmatrix} x(t) \\ y(t) \end{pmatrix} = c_1\vec{v}_1 e^{\lambda_1 t} + c_2\vec{v}_2 e^{\lambda_2 t}.$$

(b) $\lambda_1 = \lambda_2 = \lambda$, real:

$$\begin{pmatrix} x(t) \\ y(t) \end{pmatrix} = c_1\vec{v}_1 e^{\lambda t} + c_2(\vec{v}_1 t + \vec{p})e^{\lambda t},$$

where $\vec{p}$ is any nonzero vector satisfying $(A - \lambda I)\vec{p} = \vec{v}_1$.

(c) $\lambda_1 = \alpha + i\beta$, complex: write $\vec{v}_1 = \vec{u}_1 + i\vec{u}_2$, where $\vec{u}_1$ and $\vec{u}_2$ are both real vectors.

$$\begin{pmatrix} x(t) \\ y(t) \end{pmatrix} = c_1[e^{\alpha t}\cos(\beta t)\vec{u}_1 - e^{\alpha t}\sin(\beta t)\vec{u}_2] + c_2[-e^{\alpha t}\cos(\beta t)\vec{u}_2 - e^{\alpha t}\sin(\beta t)\vec{u}_1]. \tag{3.15}$$

3.4.4 Examples of the eigenvalue method

Our first example is based on a historical incident.

Example 3.4.4. In 1805, twenty-seven British ships, led by Admiral Nelson, defeated thirty-three French and Spanish ships, under French Admiral Pierre-Charles Villeneuve. See Figure 3.16.

- British fleet lost: zero,
- Franco-Spanish fleet lost: twenty-two ships.

Figure 3.16: *The Battle of Trafalgar*, by William Clarkson Stanfield (1836)

The battle is modeled by the following system of differential equations:

$$\begin{cases} x'(t) = -y(t), & x(0) = 27, \\ y'(t) = -25x(t), & y(0) = 33. \end{cases}$$

How can you use Sage to solve this?

We use the method of eigenvalues and eigenvectors: if the eigenvalues $\lambda_1 \neq \lambda_2$ of $A = \begin{pmatrix} 0 & -1 \\ -25 & 0 \end{pmatrix}$ are distinct, then the general solution can be written

$$\begin{pmatrix} x(t) \\ y(t) \end{pmatrix} = c_1 \vec{v_1} e^{\lambda_1 t} + c_2 \vec{v_2} e^{\lambda_2 t},$$

where $\vec{v_1}, \vec{v_2}$ are eigenvectors of A. Solving

$$A\vec{v} = \lambda\vec{v}$$

gives us the eigenvalues and eigenvectors. Using Sage, this is easy:

Sage
```
sage: A = matrix([[0, -1], [-25, 0]])
sage: A.eigenspaces_right()
[
(5, Vector space of degree 2 and dimension 1 over Rational Field
User basis matrix:
[ 1 -5]),
(-5, Vector space of degree 2 and dimension 1 over Rational Field
User basis matrix:
[1 5])
]
```

Therefore,

$$\lambda_1 = 5, \quad \vec{v_1} = \begin{pmatrix} 1 \\ -5 \end{pmatrix},$$

$$\lambda_2 = -5, \quad \vec{v_2} = \begin{pmatrix} 1 \\ 5 \end{pmatrix},$$

so

$$\begin{pmatrix} x(t) \\ y(t) \end{pmatrix} = c_1 \begin{pmatrix} 1 \\ -5 \end{pmatrix} e^{5t} + c_2 \begin{pmatrix} 1 \\ 5 \end{pmatrix} e^{-5t}.$$

We must solve for c_1, c_2. The initial conditions give

$$\begin{pmatrix} 27 \\ 33 \end{pmatrix} = c_1 \begin{pmatrix} 1 \\ -5 \end{pmatrix} + c_2 \begin{pmatrix} 1 \\ 5 \end{pmatrix} = \begin{pmatrix} c_1 + c_2 \\ -5c_1 + 5c_2 \end{pmatrix}$$

These equations for c_1, c_2 can be solved using Sage:

Sage
```
sage: c1,c2 = var("c1,c2")
sage: solve([c1+c2==27, -5*c1+5*c2==33],[c1,c2])
[[c1 == (51/5), c2 == (84/5)]]
```

This Sage computation gives $c_1 = 51/5$ and $c_2 = 84/5$. This gives the solution to the system of differential equations as

$$x(t) = \frac{51}{5}e^{5t} + \frac{84}{5}e^{-5t}, \quad y(t) = -51e^{5t} + 84e^{-5t}.$$

The solution satisfies

$$\begin{array}{ll} x(0.033) = 26.27\ldots & (\text{"0 losses"}), \\ y(0.033) = 11.07\ldots & (\text{"22 losses"}), \end{array}$$

consistent with the losses in the actual battle.

Example 3.4.5. Solve

$$\begin{cases} x' &= -4y, \quad x(0) = 400, \\ y' &= -x, \quad y(0) = 100, \end{cases}$$

using the eigenvalue method. (See also Example 3.3.3.)

The associated matrix is

$$A = \begin{pmatrix} 0 & -4 \\ -1 & 0 \end{pmatrix},$$

whose eigenvectors and eigenvalues are

$$\lambda_1 = 2, \quad v_1 = \begin{pmatrix} -2 \\ 1 \end{pmatrix},$$

and

$$\lambda_2 = -2, \quad v_2 = \begin{pmatrix} 2 \\ 1 \end{pmatrix}.$$

Therefore, the general solution is

$$\begin{pmatrix} x \\ y \end{pmatrix} = c_1 \begin{pmatrix} -2 \\ 1 \end{pmatrix} e^{2t} + c_2 \begin{pmatrix} 2 \\ 1 \end{pmatrix} e^{-2t}.$$

We must now solve for c_1 and c_2. The initial conditions imply

$$\begin{pmatrix} 400 \\ 100 \end{pmatrix} = c_1 \begin{pmatrix} -2 \\ 1 \end{pmatrix} + c_2 \begin{pmatrix} 2 \\ 1 \end{pmatrix} = \begin{pmatrix} -2c_1 + 2c_2 \\ c_1 + c_2 \end{pmatrix}.$$

The solution to the equations $-2c_1 + 2c_2 = 400$, $c_1 + c_2 = 100$, is (check this!) $c_1 = -50$, $c_2 = 150$. Therefore, the solution to the system of differential equations is

$$x = 100e^{2t} + 300e^{-2t}, \quad y = -50e^{2t} + 150e^{-2t}.$$

The "x-men" win.

Example 3.4.6. The eigenvalues and eigenvectors can be computed using Sage numerically as follows.

Sage

```
sage: A = matrix([[1,2],[3,4]])
sage: A.eigenvalues()
[-0.3722813232690144?, 5.372281323269015?]
sage: A.eigenvectors_right()

[(-0.3722813232690144?, [(1, -0.6861406616345072?)], 1),
 (5.372281323269015?, [(1, 2.186140661634508?)], 1)]
```

In some cases, they can be computed using Sage's "exact arithmetic" (as opposed to floating point) as follows.

Sage

```
sage: A = matrix(QQ[I],[[1,1],[-5,-1]])
sage: A.eigenvalues()
[2*I, -2*I]
sage: A.eigenvectors_right()

[(2*I, [
(1, 2*I - 1)
], 1), (-2*I, [
(1, -2*I - 1)
], 1)]
```

Example 3.4.7. Solve

$$x'(t) = x(t) - y(t), \quad y'(t) = 4x(t) + y(t), \quad x(0) = -1, \quad y(0) = 1.$$

Let

$$A = \begin{pmatrix} 1 & -1 \\ 4 & 1 \end{pmatrix},$$

and so the characteristc polynomial is

$$p(x) = \det(A - xI) = x^2 - 2x + 5.$$

The eigenvalues are

$$\lambda_1 = 1 + 2i, \quad \lambda_2 = 1 - 2i,$$

so $\alpha = 1$ and $\beta = 2$. Eigenvectors $\vec{v}_1, \vec{v}_2$ are given by

$$\vec{v}_1 = \begin{pmatrix} -1 \\ 2i \end{pmatrix}, \quad \vec{v}_2 = \begin{pmatrix} -1 \\ -2i \end{pmatrix},$$

though we actually only need to know $\vec{v}_1$. The real and imaginary parts of $\vec{v}_1$ are

$$\vec{u}_1 = \begin{pmatrix} -1 \\ 0 \end{pmatrix}, \quad \vec{u}_2 = \begin{pmatrix} 0 \\ 2 \end{pmatrix}.$$

The solution is then

$$\begin{pmatrix} x(t) \\ y(t) \end{pmatrix} = \begin{pmatrix} -c_1 e^t \cos(2t) + c_2 e^t \sin(2t) \\ -2c_1 e^t \sin(2t) - 2c_2 e^t \cos(2t) \end{pmatrix},$$

so $x(t) = -c_1 e^t \cos(2t) + c_2 e^t \sin(2t)$ and $y(t) = -2c_1 e^t \sin(2t) - 2c_2 e^t \cos(2t)$.

Since $x(0) = -1$, we solve to get $c_1 = 1$. Since $y(0) = 1$, we get $c_2 = -1/2$. The solution is: $x(t) = -e^t \cos(2t) - \frac{1}{2} e^t \sin(2t)$ and $y(t) = -2e^t \sin(2t) + e^t \cos(2t)$.

Example 3.4.8. Solve

$$x'(t) = -2x(t) + 3y(t), \quad y'(t) = -3x(t) + 4y(t).$$

Let

$$A = \begin{pmatrix} -2 & 3 \\ -3 & 4 \end{pmatrix},$$

and so the characteristic polynomial is

$$p(x) = \det(A - xI) = x^2 - 2x + 1.$$

The eigenvalues are

$$\lambda_1 = \lambda_2 = 1.$$

An eigenvector $\vec{v}_1$ is given (using (3.14)) by

$$\vec{v}_1 = \begin{pmatrix} 3 \\ 3 \end{pmatrix}.$$

Since we can multiply any eigenvector by a nonzero scalar and get another eigenvector, we shall use instead

$$\vec{v}_1 = \begin{pmatrix} 1 \\ 1 \end{pmatrix}.$$

Let $\vec{p} = \begin{pmatrix} r \\ s \end{pmatrix}$ be any nonzero vector satisfying $(A - \lambda I)\vec{p} = \vec{v}_1$. This means

$$\begin{pmatrix} -2-1 & 3 \\ -3 & 4-1 \end{pmatrix} \begin{pmatrix} r \\ s \end{pmatrix} = \begin{pmatrix} 1 \\ 1 \end{pmatrix}.$$

There are infinitely many possible solutions but we simply take $r = 0$ and $s = 1/3$, so

$$\vec{p} = \begin{pmatrix} 0 \\ 1/3 \end{pmatrix}.$$

The solution is

$$\begin{pmatrix} x(t) \\ y(t) \end{pmatrix} = c_1 \begin{pmatrix} 1 \\ 1 \end{pmatrix} e^t + c_2 \left(\begin{pmatrix} 1 \\ 1 \end{pmatrix} t + \begin{pmatrix} 0 \\ 1/3 \end{pmatrix} \right) e^t,$$

or $x(t) = c_1 e^t + c_2 t e^t$ and $y(t) = c_1 e^t + \frac{1}{3} c_2 e^t + c_2 t e^t$.

Exercises

1. Use either the Laplace transform method or the eigenvalue/eigenvector method to find the steady-state solution to the initial value problem $x' = -x - z$, $y' = -x - y$, $z' = 2x + z$, $x(0) = 0$, $y(0) = 0$, $z(0) = 2$.

2. Use Sage and the eigenvalue/eigenvector method to solve this: A battle is modeled by

$$\begin{cases} x' = -4y, & x(0) = 150, \\ y' = -x, & y(0) = 40. \end{cases}$$

(a) Write the solutions in parametric form. (b) Who wins? When? State the losses for each side.

3. All 300 x-men are fighting all 101 y-men, the battle being modeled by

$$\begin{cases} x' = -9y, & x(0) = 300, \\ y' = -x, & y(0) = 101. \end{cases}$$

Solve this system using the eigenvalue/eigenvector method. Who wins? When?

4. All 60 x-men are fighting all 15 y-men, the battle being modeled by

$$\begin{cases} x' = -4y, & x(0) = 60, \\ y' = -x, & y(0) = 15. \end{cases}$$

Solve this system using the eigenvalue/eigenvector method. Who wins? When?

3.5 Introduction to variation of parameters for systems

The method called variation of parameters for systems of ordinary differential equations has no relation to the method of variation of parameters for second-order ordinary differential equations discussed in §2.6 except for the name.

3.5.1 Motivation

Recall that when we solved the first-order ordinary differential equation

$$y' = ay, \qquad y(0) = y_0, \tag{3.16}$$

for $y = y(t)$ using the method of separation of variables, we got the formula

$$y = ce^{at} = e^{at}c, \tag{3.17}$$

where c is a constant depending on the initial condition (in fact, $c = y(0)$).

Consider instead a 2×2 system of linear first-order ordinary differential equations in the form

$$\begin{cases} x' = ax + by, & x(0) = x_0, \\ y' = cx + dy, & y(0) = y_0. \end{cases}$$

This can be rewritten in the form

$$\vec{X}' = A\vec{X}, \tag{3.18}$$

where $\vec{X} = \vec{X}(t) = \begin{pmatrix} x(t) \\ y(t) \end{pmatrix}$, and A is the matrix

$$A = \begin{pmatrix} a & b \\ c & d \end{pmatrix}.$$

We can solve (3.18) analogously to (3.16), to obtain

$$\vec{X} = e^{tA}\vec{c}. \tag{3.19}$$

Here,

$$\vec{c} = \begin{pmatrix} x(0) \\ y(0) \end{pmatrix}$$

is a constant depending on the initial conditions, and e^{tA} is a matrix exponential defined by

$$e^B = \sum_{n=0}^{\infty} \frac{1}{n!} B^n,$$

for any square matrix B (yes, this series does actually converge!). You might be thinking to yourself: I can't compute the matrix exponential so what good is this formula (3.19)? That is a good question, and the answer is that the eigenvalues and eigenvectors of the matrix A enable you to compute e^{tA}. This is the basis for the formulas for the solution of a system of ordinary differential equations using the eigenvalue method discussed in §3.4. None the less, we shall instead use a matrix called the "fundamental matrix" (defined in the next paragraph[8]) in place of e^{tA}.

The eigenvalue method in §3.4 shows us how to write every solution to (3.18) in the form

$$\vec{X} = c_1\vec{X}_1(t) + c_2\vec{X}_2(t),$$

for some vector-valued solutions $\vec{X}_1(t)$ and $\vec{X}_2(t)$, called *fundamental solutions*, which can be simply expressed in terms of the eigenvalues and eigenvectors of A. Frequently, we call the matrix of fundamental solutions,

$$\Phi = \left(\vec{X}_1(t), \vec{X}_2(t)\right), \tag{3.20}$$

---

[8] In fact, if the $\vec{X}_i$'s in (3.20) are carefully selected then $\Phi = e^{tA}$. This is not obvious and is beyond the scope of this text.

the *fundamental matrix*, where we are regarding $\vec{X}_1(t), \vec{X}_2(t)$ as *column* vectors. The fundamental matrix is, roughly speaking, e^{tA}. It is analogous to the Wronskian of two fundamental solutions to a second order ordinary differential equation.

See the examples below for more details.

3.5.2 The method

Recall that when we solved the first-order ordinary differential equation

$$y' + p(t)y = q(t) \tag{3.21}$$

for $y = y(t)$ using the method of integrating factors, we got the formula

$$y = (e^{\int p(t)\,dt})^{-1}\left(\int e^{\int p(t)\,dt} q(t)\,dt \; + \; c\right). \tag{3.22}$$

Consider a 2×2 system of linear first-order ordinary differential equations in the form

$$\begin{cases} x' = ax + by + f(t), & x(0) = x_0, \\ y' = cx + dy + g(t), & y(0) = y_0. \end{cases}$$

This can be rewritten in the form

$$\vec{X}' = A\vec{X} + \vec{F}, \tag{3.23}$$

where $\vec{F} = \vec{F}(t) = \begin{pmatrix} f(t) \\ g(t) \end{pmatrix}$. Equation (3.23) can be seen to be in a form analogous to (3.21) by replacing $\vec{X}$ by y, A by $-p$, and $\vec{F}$ by q. It turns out that (3.23) can be solved in a way analogous to (3.21) as well. Here is the *variation of parameters formula for systems*:

$$\vec{X} = \Phi \cdot \left(\int \Phi^{-1}\vec{F}(t)\,dt \; + \; \vec{c}\right), \tag{3.24}$$

where $\vec{c} = \begin{pmatrix} c_1 \\ c_2 \end{pmatrix}$ is a constant vector determined by the initial conditions and Φ is the fundamental matrix.

Example 3.5.1. A battle between x-men and y-men is modeled by

$$\begin{cases} x' = -y + 1, & x(0) = 100, \\ y' = -4x + e^t, & y(0) = 50. \end{cases}$$

The nonhomogeneous terms 1 and e^t represent reinforcements. Find out who wins, when, and the number of survivors.

Here A is the matrix

$$A = \begin{pmatrix} 0 & -1 \\ -4 & 0 \end{pmatrix}$$

and $\vec{F} = \vec{F}(t) = \begin{pmatrix} 1 \\ e^t \end{pmatrix}$.

In the method of variation of parameters, you must solve the homogeneous system first.

The eigenvalues of A are $\lambda_1 = 2$, $\lambda_2 = -2$, with associated eigenvectors $\vec{v}_1 = \begin{pmatrix} 1 \\ -2 \end{pmatrix}$, $\vec{v}_2 = \begin{pmatrix} 1 \\ 2 \end{pmatrix}$, respectively. The general solution to the homogeneous system

$$\begin{cases} x' = -y, \\ y' = -4x \end{cases}$$

is

$$\vec{X} = c_1 \begin{pmatrix} 1 \\ -2 \end{pmatrix} e^{2t} + c_2 \begin{pmatrix} 1 \\ 2 \end{pmatrix} e^{-2t} = c_1 \vec{X}_1(t) + c_2 \vec{X}_2(t),$$

where

$$\vec{X}_1(t) = \begin{pmatrix} e^{2t} \\ -2e^{2t} \end{pmatrix}, \quad \vec{X}_2(t) = \begin{pmatrix} e^{-2t} \\ 2e^{-2t} \end{pmatrix}.$$

For the solution of the nonhomogeneous equation, we must compute the fundamental matrix:

$$\Phi = \begin{pmatrix} e^{2t} & e^{-2t} \\ -2e^{2t} & 2e^{-2t} \end{pmatrix}, \quad \text{so} \quad \Phi^{-1} = \frac{1}{4} \begin{pmatrix} 2e^{-2t} & -e^{-2t} \\ 2e^{2t} & e^{2t} \end{pmatrix}.$$

Next, we compute the product

$$\Phi^{-1}\vec{F} = \frac{1}{4} \begin{pmatrix} 2e^{-2t} & -e^{-2t} \\ 2e^{2t} & e^{2t} \end{pmatrix} \begin{pmatrix} 1 \\ e^t \end{pmatrix} = \begin{pmatrix} \frac{1}{2}e^{-2t} - \frac{1}{4}e^{-t} \\ \frac{1}{2}e^{2t} + \frac{1}{4}e^{3t} \end{pmatrix}$$

and its integral

$$\int \Phi^{-1}\vec{F}\, dt = \begin{pmatrix} -\frac{1}{4}e^{-2t} + \frac{1}{4}e^{-t} \\ \frac{1}{4}e^{2t} + \frac{1}{12}e^{3t} \end{pmatrix}.$$

Finally, to finish (3.24), we compute the matrix times vector product,

$$\begin{aligned} \Phi(\textstyle\int \Phi^{-1}\vec{F}(t)\, dt \ + \ \vec{c}) &= \begin{pmatrix} e^{-2t} & e^{-2t} \\ -2e^{2t} & 2e^{2t} \end{pmatrix} \begin{pmatrix} -\frac{1}{4}e^{-2t} + \frac{1}{4}e^{-t} + c1 \\ \frac{1}{4}e^{2t} + \frac{1}{12}e^{3t} + c_2 \end{pmatrix} \\ &= \begin{pmatrix} c_1 e^{2t} + \frac{1}{3}e^t + c_2 e^{-2t} \\ 1 - \frac{1}{3}e^t - 2c_1 e^{2t} + 2c_2 e^{-2t} \end{pmatrix}. \end{aligned}$$

This gives the general solution to the original system

$$x(t) = c_1 e^{2t} + \frac{1}{3}e^t + c_2 e^{-2t}$$

and

$$y(t) = 1 - \frac{1}{3}e^t - 2c_1 e^{2t} + 2c_2 e^{-2t}.$$

We aren't done! It remains to compute c_1, c_2 using the initial conditions. For this, solve

$$\frac{1}{3} + c_1 + c_2 = 100, \quad \frac{2}{3} - 2c_1 + 2c_2 = 50.$$

We get

$$c_1 = 75/2, \quad c_2 = 373/6,$$

so

$$x(t) = \frac{75}{2}e^{2t} + \frac{1}{3}e^t + \frac{373}{6}e^{-2t}$$

and

$$y(t) = 1 - \frac{1}{3}e^t - 75e^{2t} + \frac{373}{3}e^{-2t}.$$

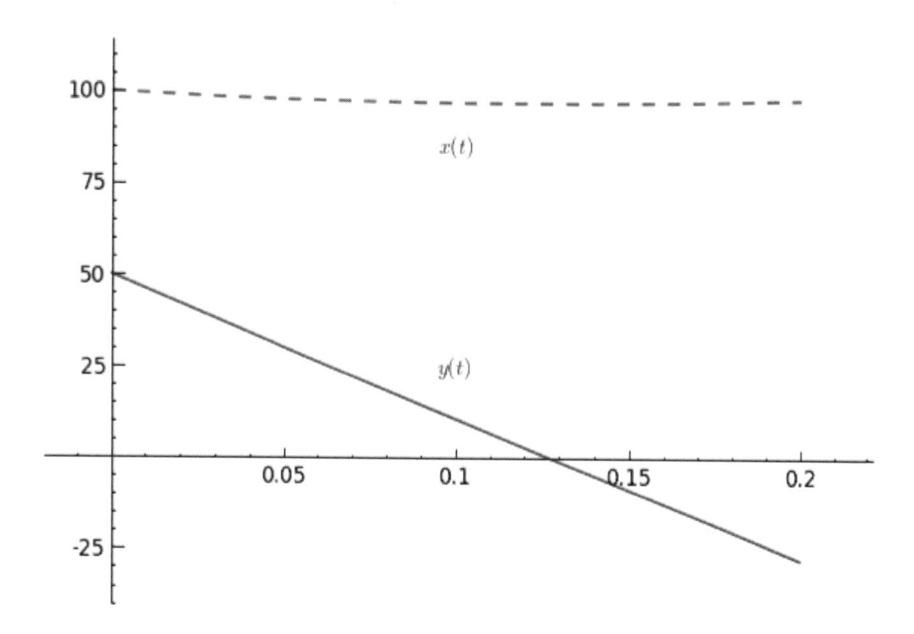

Figure 3.17: Solution to system $x' = -y + 1$, $x(0) = 100$, $y' = -4x + e^t$, $y(0) = 50$.

As you can see from Figure 3.17, the x-men win. The solution to $y(t) = 0$ is about $t_0 = 0.1279774\ldots$ and $x(t_0) = 96.9458\ldots$ "survive."

Example 3.5.2. Solve

$$\begin{cases} x' = -y + 1, \\ y' = x + \cot(t). \end{cases}$$

Here A is the matrix

$$A = \begin{pmatrix} 0 & -1 \\ 1 & 0 \end{pmatrix}$$

and $\vec{F} = \vec{F}(t) = \begin{pmatrix} 0 \\ \cot(t) \end{pmatrix}$.

In the method of variation of parameters, you must solve the homogeneous system first.

The eigenvalues of A are $\lambda_1 = i$, $\lambda_2 = -i$, with associated eigenvectors $\vec{v}_1 = \begin{pmatrix} 1 \\ -i \end{pmatrix}$, $\vec{v}_2 = \begin{pmatrix} 1 \\ i \end{pmatrix}$, respectively. Therefore, in the notation of (3.15), we have $\alpha = 0$, $\beta = 1$, $\vec{u}_1 = \begin{pmatrix} 1 \\ 0 \end{pmatrix}$, and $\vec{u}_2 = \begin{pmatrix} 0 \\ -1 \end{pmatrix}$.

The general solution to the homogeneous system

$$\begin{cases} x' = -y, \\ y' = x, \end{cases}$$

is

$$\vec{X} = c_1 \left[\cos(t) \begin{pmatrix} 1 \\ 0 \end{pmatrix} - \sin(t) \begin{pmatrix} 0 \\ -1 \end{pmatrix}\right] + c_2 \left[\cos(t) \begin{pmatrix} 0 \\ -1 \end{pmatrix} + \sin(t) \begin{pmatrix} 1 \\ 0 \end{pmatrix}\right] = c_1 \vec{X}_1(t) + c_2 \vec{X}_2(t),$$

where

$$\vec{X}_1(t) = \begin{pmatrix} \cos(t) \\ \sin(t) \end{pmatrix}, \quad \vec{X}_2(t) = \begin{pmatrix} \sin(t) \\ -\cos(t) \end{pmatrix}.$$

For the solution of the nonhomogeneous equation, we must compute the fundamental matrix

$$\Phi = \begin{pmatrix} \cos(t) & \sin(t) \\ \sin(t) & -\cos(t) \end{pmatrix}, \quad \text{so} \quad \Phi^{-1} = \begin{pmatrix} \cos(t) & \sin(t) \\ \sin(t) & -\cos(t) \end{pmatrix} = \Phi,$$

since $\cos(t)^2 + \sin(t)^2 = 1$. Next, we compute the product

$$\Phi^{-1}\vec{F} = \begin{pmatrix} \cos(t) & \sin(t) \\ \sin(t) & -\cos(t) \end{pmatrix} \begin{pmatrix} 0 \\ \cot(t) \end{pmatrix} = \begin{pmatrix} \cos(t) \\ -\cos(t)^2/\sin(t) \end{pmatrix} = \begin{pmatrix} \cos(t) \\ \sin(t) - 1/\sin(t) \end{pmatrix}$$

and its integral

$$\int \Phi^{-1}\vec{F}\,dt = \begin{pmatrix} \sin(t) \\ -\cos(t) - \frac{1}{2}\frac{\cos(t)-1}{\cos(t)+1} \end{pmatrix}.$$

Finally, we compute

$$\Phi(\int \Phi^{-1}\vec{F}(t)\,dt\ +\ \vec{c}) = \begin{pmatrix} \cos(t) & \sin(t) \\ \sin(t) & -\cos(t) \end{pmatrix} \left[\begin{pmatrix} c_1 \\ c_2 \end{pmatrix} + \begin{pmatrix} \sin(t) \\ -\cos(t) - \frac{1}{2}\frac{\cos(t)-1}{\cos(t)+1} \end{pmatrix} \right]$$
$$= \begin{pmatrix} c_1 \cos(t) + c_2 \sin(t) - \frac{1}{2}\frac{\sin(t)\cos(t)}{(\cos(t)+1)} + \frac{1}{2}\frac{\sin(t)}{(\cos(t)+1)} \\ c_1 \sin(t) - c_2 \cos(t) + 1 + \frac{1}{2}\frac{\cos(t)^2}{(\cos(t)+1)} - \frac{1}{2}\frac{\cos(t)}{(\cos(t)+1)} \end{pmatrix}.$$

Therefore,

$$x(t) = c_1 \cos(t) + c_2 \sin(t) - \frac{1}{2}\frac{\sin(t)\cos(t)}{(\cos(t)+1)} + \frac{1}{2}\frac{\sin(t)}{(\cos(t)+1)}$$

and

$$y(t) = c_1 \sin(t) - c_2 \cos(t) + 1 + \frac{1}{2}\frac{\cos(t)^2}{(\cos(t)+1)} - \frac{1}{2}\frac{\cos(t)}{(\cos(t)+1)}.$$

Exercises

1. Use Sage and the method of variation of parameters to solve this: A battle is modeled by
$$\begin{cases} x' = -4y, & x(0) = 150, \\ y' = -x + 1, & y(0) = 40. \end{cases}$$
(a) Write the solutions in parametric form. (b) Who wins?

2. All 300 x-men are fighting all 101 y-men, the battle being modeled by
$$\begin{cases} x' = -9y + 1, & x(0) = 300, \\ y' = -x - 1, & y(0) = 101. \end{cases}$$
Solve this system using the method of variation of parameters. Who wins?

3. All 60 x-men are fighting all 15 y-men, the battle being modeled by
$$\begin{cases} x' = -4y - 1, & x(0) = 60, \\ y' = -x + 1, & y(0) = 15. \end{cases}$$
Solve this system using the method of variation of parameters. Who wins?

3.6 Nonlinear systems

Most differential equations that people are interested in are nonlinear. Unfortunately these equations are usually not solvable (by functions whose values can be determined exactly), so we must rely on a combination of numerical approximations to the solutions and methods for analyzing some aspect of the solutions. These latter methods are often lumped together under the name "qualitative methods" and they are part of the modern mathematical discipline of "dynamical systems."

In this section we will introduce some of these methods for first-order nonlinear systems. Recall that a higher-order nonlinear ODE can always be converted to a first-order system (see §1.6.3), so there is no loss of generality.

3.6.1 Linearizing near equilibria

One of the simplest and most fruitful things we can do with a nonlinear system is find its equilibria—values where all of the first-order derivatives are zero—and study a linear approximation of the system near these points.

Our first example of the use of this technique will be the predator-prey system of Lotka and Volterra (the Lotka-Volterra equations). We are modeling two species, one of which (population y) preys upon the other (population x). The rate of predation is assumed to be proportional to the product xy (this is similar to the mass-action law in chemical kinetic modeling). As the predators cannot live without any prey, they are subjected to a death rate proportional to their population size. The prey are assumed to grow without predation at a rate proportional to their population size:

$$\begin{aligned} x' &= ax - bxy, \\ y' &= -cy + dxy, \end{aligned}$$

where a, b, c, and d are positive parameters.

We can find the equilibria by factoring each of the right-hand sides of this system: we need $x(a - by) = 0$ and $y(-c + dx) = 0$. The two solutions are $(x, y) = (0, 0)$ and $(x, y) = (c/d, a/b)$. Now we linearize around each of these equilibria.

To linearize the system we compute the Jacobian matrix—the matrix of all first-order derivatives of the slope functions—and use the value of this matrix at the point of linearization as our constant matrix for a linear system. For the Lotka-Volterra system this matrix is

$$\begin{pmatrix} \frac{\partial(ax-bxy)}{\partial x} & \frac{\partial(ax-bxy)}{\partial y} \\ \frac{\partial(-cy+dxy)}{\partial x} & \frac{\partial(-cy+dxy)}{\partial y} \end{pmatrix} = \begin{pmatrix} a - by & -bx \\ dy & -c + dx \end{pmatrix},$$

and its value at $(0, 0)$ is

$$\begin{pmatrix} a & 0 \\ 0 & -c \end{pmatrix}.$$

The linearized system around the origin is

$$\begin{pmatrix} u \\ v \end{pmatrix}' = \begin{pmatrix} a & 0 \\ 0 & -c \end{pmatrix} \begin{pmatrix} u \\ v \end{pmatrix}.$$

Since this is a diagonal matrix we know the eigenvalues are simply a and $-c$, so it is unstable (more precisely, it's a saddle point).

If we evaluate the Jacobian matrix at the other equilbrium, $(c/d, a/b)$, we obtain the linearization

$$\begin{pmatrix} u \\ v \end{pmatrix}' = \begin{pmatrix} 0 & -bc/d \\ da/b & 0 \end{pmatrix} \begin{pmatrix} u \\ v \end{pmatrix},$$

where we interpet the variables u and v as approximations to the deviation from the equilibrium, that is, for small u and v we should have $u \approx x - c/d$ and $v \approx y - a/b$. The eigenvalues of this system are given by

$$\det\left(\lambda I - \begin{pmatrix} 0 & -bc/d \\ da/b & 0 \end{pmatrix}\right) = \lambda^2 + ac = 0,$$

so $\lambda = \pm\sqrt{ac}i$ and they are purely imaginary.

Trajectories for the particular case $a = b = c = d = 1$ are illustrated in Figure 3.18. Note that close to the equilibrium at $(1, 1)$ the trajectories are close to the elliptic solutions of the linearized equations for that point, but closer to the other equilibrium at the origin the solutions appear more and more hyperbolic.

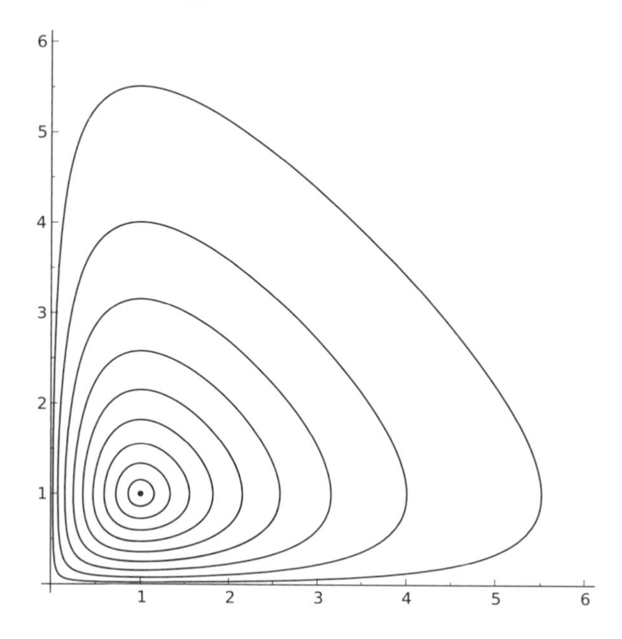

Figure 3.18: Lotka-Volterra model

Exercises

1. Find the equilibrium and compute the linearized system around it for the system $x_1' = x_1 - x_2^2$, $x_2' = x_1 + 2x_1^2 - 2x_2$.

2. Find all of the equilibria, linearizations around them, and the corresponding eigenvalues for the system $x_1' = (1 + x_1)(x_1 - x_2)$, $x_2' = (2 - x_1)(x_1 + x_2)$.

3.6.2 The nonlinear pendulum

Consider an idealized pendulum with a massless rigid shaft of length L and a small sphere of mass m swinging from a point. At first we will assume there is no friction.

When the pendulum has swung through an angle of θ from its bottom position, the component of the gravitational acceleration $-g$ in the tangential direction is $-g\sin(\theta)$. The angular acceleration is $g\sin(\theta)/L$, so the angle of the pendulum satisfies the ODE

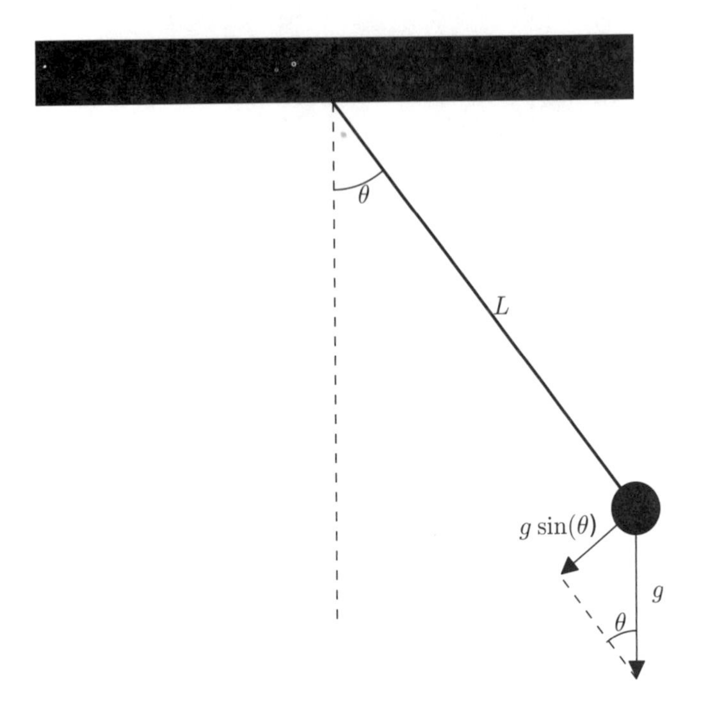

Figure 3.19: The nonlinear pendulum

$$\theta'' = -\frac{g}{L}\sin(\theta).$$

First we rewrite the ODE as a system of first-order equations by introducing the velocity $v = \theta'$ as a variable in its own right. Then we have

$$\begin{pmatrix} \theta' \\ v' \end{pmatrix} = \begin{pmatrix} v \\ -\frac{g}{L}\sin(\theta) \end{pmatrix}.$$

The equilibria occur when $\sin(\theta) = 0$ and $v = 0$, so $\theta = n\pi$ for some integer n. To linearize the system around these equilibria we compute the Jacobian matrix

$$J = \begin{pmatrix} \frac{dv}{d\theta} & \frac{dv}{dv} \\ \frac{d}{d\theta}(-\frac{g}{L}\sin(\theta)) & \frac{d}{dv}(-\frac{g}{L}\sin(\theta)) \end{pmatrix} = \begin{pmatrix} 0 & 1 \\ -\frac{g}{L}\cos(\theta) & 0 \end{pmatrix}.$$

At the equilibria, the only thing that matters when evaluating this matrix is whether θ is an even or odd multiple of π, since that determines the sign of $\cos(\theta)$:

$$\begin{pmatrix} \theta' \\ v' \end{pmatrix} = \begin{pmatrix} 0 & 1 \\ \pm\frac{g}{L} & 0 \end{pmatrix}\begin{pmatrix} \theta \\ v \end{pmatrix}.$$

When $n\pi$ is even, $\cos(\theta) = 1$, the eigenvalues of the linearized system satisfy $\lambda^2 + \frac{g}{L} = 0$, and so $\lambda = \pm\sqrt{\frac{g}{L}}i$. These are pure imaginary eigenvalues, so the linearized system has solutions which orbit the equilibria on elliptic trajectories.

When $n\pi$ is odd, $\cos(\theta) = 1$, the eigenvalues of the linearized system satisfy $\lambda^2 - \frac{g}{L} = 0$, and so $\lambda = \pm\sqrt{\frac{g}{L}}$. This pair of positive and negative real eigenvalues means that the linearized system has an unstable equilbrium, a saddle point.

3.6.3 The Lorenz equations

The Lorenz equations are a famous example of a nonlinear system in which the trajectories are *chaotic*. Roughly speaking, this means that although the solutions are close to periodic, two solutions starting close together will diverge at an exponential rate, so that after a little while they will have quite different behavior.

$$\begin{aligned} x' &= \sigma(y - x), \\ y' &= x(\rho - z) - y, \\ z' &= xy - \beta z. \end{aligned}$$

Here σ, β, and ρ are positive parameters.

Edward Lorenz introduced these well-known equations[9] in 1963 as a very simplified model of convection in the atmosphere. This model is in turn derived from a twelve-variable model whose variables are more closely related to physical quantities.

This system has a wide variety of interesting behaviors for different parameter values. The most commonly studied case is when $\sigma = 10$, $\beta = 8/3$, and $r = 28$. A plot of the x value is given in Figure 3.20 for this case for two initial conditions very close together ($x_0 = 5$ and $x_0 = 5.001$).

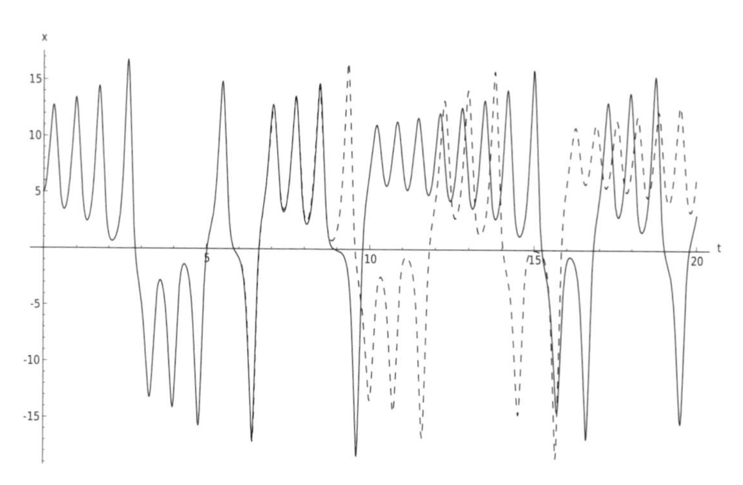

Figure 3.20: The Lorenz model, x versus t

[9] In Wikipedia, if you search for "Lorenz model" and look for the equations describing this model, you will find these.

In the (x, y) plane it is easier to see how a typical trajectory slowly alternates between oscillating around two equilibria (Figure 3.21).

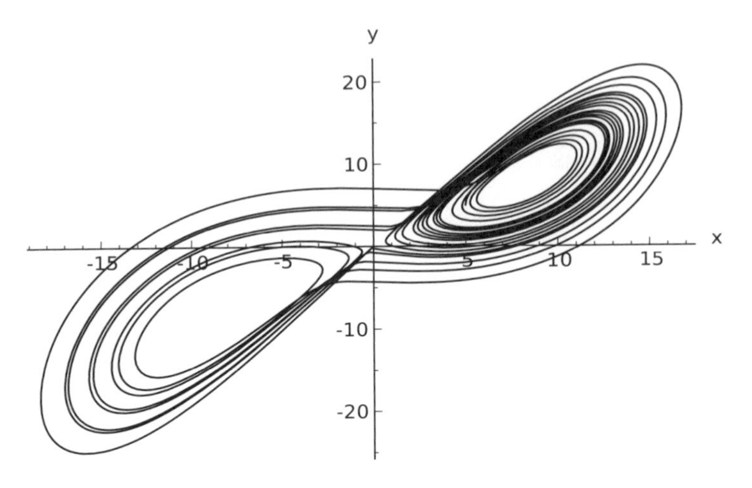

Figure 3.21: The Lorenz model, x versus y

Exercises

1. In 1958, Tsuneji Rikitake formulated a simple model of the Earth's magnetic core to explain the oscillations in the polarity of the magnetic field [Rikitake]. The equations for his model are

$$\begin{aligned} x' &= -\mu x + yz, \\ y' &= -\mu y + (z - a)x, \\ z' &= 1 - xy, \end{aligned}$$

where a and μ are positive constants. Find the equilibria for this system for $a = \mu = 1$, and compute the eigenvalues of the linearized system at those equilibria.

3.6.4 Zombies attack

Mathematical research does not excite the public imagination very often, but in 2009 a paper called "When zombies attack! Mathematical modelling of an outbreak of zombie infection" [M-zom] was published that received much attention. The mathematical model the authors use is based on a "serious" model used to investigate infectious diseases. The zombie infection pattern they assume follows the one presented by George Romero in his famous (and now public-domain) 1968 movie *Night of the Living Dead.*

Let

- S represent people (the "susceptibles"),

- Z be the number of zombies, and
- R (the "removed") represent (a) deceased zombies, (b) bitten people (who are sometimes turned into zombies), or (c) dead people.

Figure 3.22: Zombies in George Romero's *Night of the Living Dead*

The simplest system of ODEs developed in that paper is

$$\begin{aligned} S' &= B - \beta SZ - \delta S, \\ Z' &= \beta SZ + \zeta R - \alpha SZ, \\ R' &= \delta S + \alpha SZ - \zeta R. \end{aligned}$$

What do these terms mean and how do they arise? We assume the following:

- People (counted by S) have a constant birth rate B.
- A proportion δ of people die a non-zombie-related death. (This accounts for the $-\delta S$ term in the S' line and the $+\delta S$ term in the R' line.)
- A proportion ζ of dead humans (counted by R) can resurrect and become zombies. (This accounts for the $+\zeta R$ term in the Z' line and the $-\zeta R$ term in the R' line.)

There are nonlinear terms as well. These correspond to interactions between a person and a zombie—the "SZ terms." We assume:

- Some of these interactions result in a person killing a zombie by destroying its brain. (This accounts for the $-\alpha SZ$ term in the Z' line and the $+\alpha SZ$ term in the R' line.)

- Some of these interactions results in a zombie infecting a person and turning that person into a zombie.
  (This accounts for the $-\beta SZ$ term in the S' line and the $+\beta SZ$ term in the Z' line.)

These nonlinear terms mean that the system is too complicated to solve by simple methods, such as the method of eigenvalues or Laplace transforms.

However, solutions to the Zombies Attack model can be numerically approximated in Sage.
Let's take

$$\alpha = 0.005, \quad \beta = 0.004, \quad \zeta = 0.0001,$$
$$\delta = 0.0001, \quad B = 0,$$

and solve the above system numerically. What do we get?

Sage

```
sage: from sage.calculus.desolvers import desolve_system_rk4
sage: t,s,z,r = var('t,s,z,r')
sage: a,b,zeta,d,B = 0.005,0.004,0.0001,0.0001,0.0
sage: P = desolve_system_rk4([B-b*s*z-d*s,b*s*z-zeta*r-a*s*z,d*s+a*s*z-zeta*r],
 [s,z,r],ics=[0,11,10,5] ,ivar=t,end_points=30)
sage: Ps = list_plot([[t,s] for t,s,z,r in P],plotjoined=True,
 legend_label='People')
sage: Pz = list_plot([[t,z] for t,s,z,r in P],plotjoined=True,rgbcolor='red',
 legend_label='Zombies')
sage: Pr = list_plot([[t,r] for t,s,z,r in P],plotjoined=True,rgbcolor='black',
 legend_label='Deceased')
sage: show(Ps+Pz+Pr)
```

The plot of the new solution is given in Figure 3.23. The zombies win.

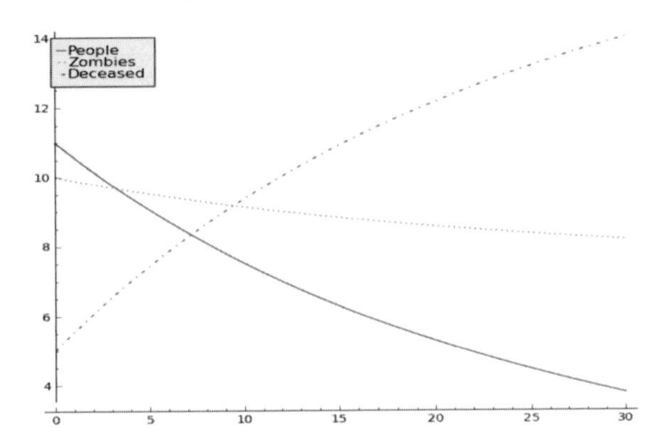

Figure 3.23: A suseptibles, zombies and removed plot.

Too many zombies are infecting people (this is the constant β, which is relatively large).

Let's make β smaller

$$\alpha = 0.005, \quad \beta = 0.002, \quad \zeta = 0.0001,$$

$$\delta = 0.0001, \quad B = 0,$$

and solve the above system numerically. Now, what do we get?

Sage

```
sage: from sage.calculus.desolvers import desolve_system_rk4
sage: t,s,z,r = var('t,s,z,r')
sage: a,b,zeta,d,B = 0.005,0.002,0.0001,0.0001,0.0
sage: P = desolve_system_rk4([B-b*s*z-d*s,b*s*z-zeta*r-a*s*z,d*s+a*s*z-zeta*r],
 [s,z,r],ics=[0,11,10,5] ,ivar=t,end_points=30)
sage: Ps = list_plot([[t,s] for t,s,z,r in P],plotjoined=True,
 legend_label='People')
sage: Pz = list_plot([[t,z] for t,s,z,r in P],plotjoined=True,rgbcolor='red',
 legend_label='Zombies')
sage: Pr = list_plot([[t,r] for t,s,z,r in P],plotjoined=True,rgbcolor='black',
 legend_label='Deceased')
sage: show(Ps+Pz+Pr)
```

The plot of the solutions is given Figure 3.24. Finally, the zombies lose.

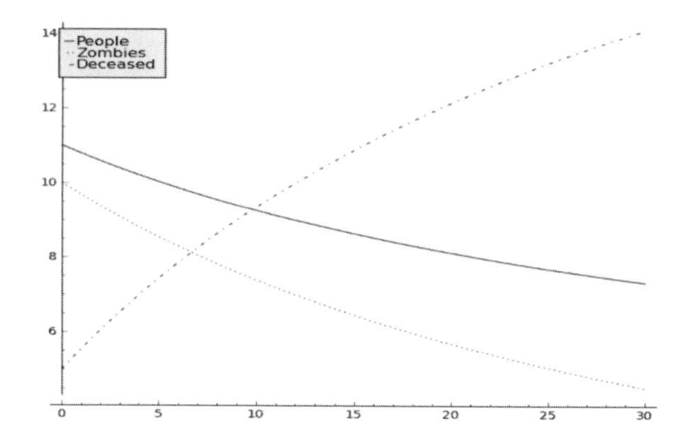

Figure 3.24: Another suseptibles, zombies, and removed plot

The moral of the story: to survive, you must *protect yourself from zombies.*

You may wonder about the importance of learning about this model. Although it is humorous, it is also very closely related to many models of the transmission of infectious diseases called *SIR models* (the initials stand for susceptible, infected, and recovered). These models can be very important for developing efficient public health strategies.

In the zombie model, the terms proportional to SZ represent interactions between people and zombies, and these are the only nonlinear terms. This and the SIR models are similar to the Lotka-Volterra model of predators and prey in that they model the interaction frequency of two groups by a quadratic term.

Example 3.6.1. Solutions to the Zombies Attack model can be numerically approximated in Sage.

Sage

```
sage: from sage.calculus.desolvers import desolve_system_rk4
sage: t,s,z,r = var('t,s,z,r')
sage: a,b,zeta,d,B = 0.005,0.0095,0.0001,0.0001,0.0
sage: P = desolve_system_rk4([B-b*s*z-d*s,b*s*z-zeta*r-a*s*z,d*s+a*s*z-zeta*r],
 [s,z,r],ics=[0,11,10,5] ,ivar=t,end_points=30)
sage: Ps = list_plot([[t,s] for t,s,z,r in P],plotjoined=True,
 legend_label='People')
sage: Pz = list_plot([[t,z] for t,s,z,r in P],plotjoined=True,rgbcolor='red',
 legend_label='Zombies')
sage: Pr = list_plot([[t,r] for t,s,z,r in P],plotjoined=True,rgbcolor='black',
 legend_label='Deceased')
sage: show(Ps+Pz+Pr)
```

This is displayed in Figure 3.25.

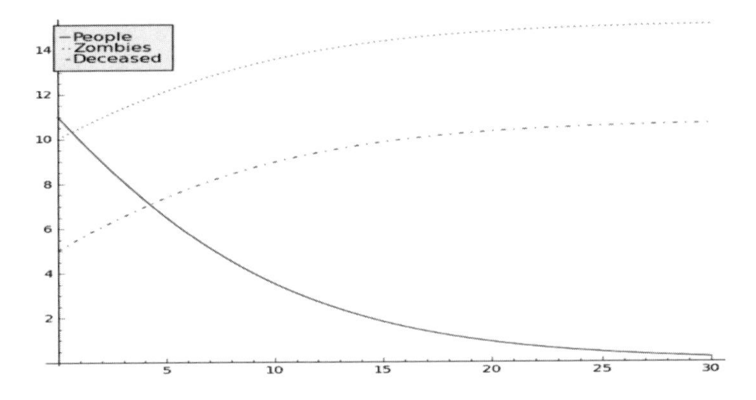

Figure 3.25: Zombies Attack model.

Exercises

1*. Introduce a new term in the zombie attack equations representing fights between zombies. What effect would this have on the equilibria and their stability for parameter values $\alpha = 0.005$, $\beta = 0.0095$, $\zeta = 0.0001$, and $\delta = 0.0001$?

Chapter 4

Introduction to partial differential equations

> The deep study of nature is the most fruitful source of mathematical discoveries.
>
> —*Jean-Baptist-Joseph Fourier*

4.1 Introduction to separation of variables

Recall that a *partial differential equation* (PDE) is an equation satisfied by an unknown function (called the dependent variable) and its partial derivatives. The variables you differentiate with respect to are called the independent variables. If there is only one independent variable then it is called an *ordinary differential equation.*

Examples include

- the Laplace equation $\frac{\partial^2 u}{\partial x^2} + \frac{\partial^2 u}{\partial y^2} = 0$, where u is the dependent variable and x, y are the independent variables,
- the heat equation $u_t = \alpha u_{xx}$,
- and the wave equation $u_{tt} = c^2 u_{xx}$.

All these PDEs are of second order (you have to differentiate twice to express the equation). Here, we consider a first-order PDE which arises in applications and use it to introduce the method of solution called *separation of variables.*

At the present time, `Sage` has limited tools for solving PDEs symbolically. (In all fairness, this is a very hard problem and not many computer algebra systems have methods for solving PDEs symbolically.) There are some tools for solving PDEs numerically, but those techniques go well beyond the scope of this book. We recommend that the interested reader look into `clawpack` (which has started some integration with `Sage`) and `FEMhub`[1], which uses `Sympy` and other packages also used by `Sage`

4.1.1 The transport or advection equation

Advection is the transport of some conserved scalar quantity in a vector field. A good example is the transport of pollutants or silt in a river (the motion of the water carries these impurities downstream) or traffic flow.

The advection equation is the PDE governing the motion of a conserved quantity as it is advected by a given velocity field. The advection equation expressed mathematically is

$$\frac{\partial u}{\partial t} + \nabla \cdot (u\mathbf{a}) = 0,$$

where $\nabla\cdot$ is the divergence and $\mathbf{a}$ is the velocity field of the fluid. Frequently, it is assumed that $\nabla \cdot \mathbf{a} = 0$ (this is expressed by saying that *the velocity field is solenoidal*). In this case, the above equation reduces to

$$\frac{\partial u}{\partial t} + \mathbf{a} \cdot \nabla u = 0.$$

Assume we have a horizontal pipe in which water is flowing at a constant rate c in the positive x direction. Add some salt to this water and let $u(x,t)$ denote the concentration (say in lb/gal) at time t. Note that the amount of salt in an interval I of the pipe is $\int_I u(x,t)\,dx$. This concentration satisfies the *transport* (or *advection*) equation

$$u_t + cu_x = 0.$$

(For a derivation of this, see, for example, Strauss [S-pde], §1.3.) How do we solve this?

- *Solution 1*: D'Alembert noticed that the directional derivative of $u(x,t)$ in the direction $\vec{v} = \frac{1}{\sqrt{1+c^2}}\langle c, 1\rangle$ is $D_{\vec{v}}(u) = \frac{1}{\sqrt{1+c^2}}(cu_x + u_t) = 0$. Therefore, $u(x,t)$ is constant along the lines in the direction of $\vec{v}$, and so $u(x,t) = f(x-ct)$, for some function f. We will not use this method of solution in the example below but it does help visualize the shape of the solution. For instance, imagine the plot of $z = f(x-ct)$ in (x,t,z) space. The contour lying above the line $x = ct + k$ (k fixed) is the line of constant height $z = f(k)$.

- *Solution 2*: The method of "separation of variables" indicates that we start by assuming that $u(x,t)$ can be factored:

$$u(x,t) = X(x)T(t),$$

[1]This is also available online at `http://lab.femhub.org/`. Go to `http://femhub.org/` or `http://en.wikipedia.org/wiki/FEMhub_Project` for more information.

for some (unknown) functions X and T. (We shall work on removing this assumption later. The assumption "works" because partial differentiation of functions like x^2t^3 is so much simpler than partial differentiation of "mixed" functions like $\sin(x^2+t^3)$.) Substituting this into the PDE gives

$$X(x)T'(t) + cX'(x)T(t) = 0.$$

Now separate all the x's on one side and the t's on the other (divide by $X(x)T(t)$):

$$\frac{T'(t)}{T(t)} = -c\frac{X'(x)}{X(x)}.$$

(This same "trick" works when you apply the separation of variables method to other linear PDEs, such as the heat equation or wave equation, as we will see in later sections.) It is impossible for a function of an independent variable x to be identically equal to a function of an independent variable t unless both are constant. (Indeed, try taking the partial derivative of $\frac{T'(t)}{T(t)}$ with respect to x. You get 0 since it doesn't depend on x. Thus, the partial derivative of $-c\frac{X'(x)}{X(x)}$ is akso 0, so $\frac{X'(x)}{X(x)}$ is a constant!) Therefore, $\frac{T'(t)}{T(t)} = -c\frac{X'(x)}{X(x)} = K$, for some (unknown) constant K. So we have two ODEs:

$$\frac{T'(t)}{T(t)} = K, \quad \frac{X'(x)}{X(x)} = -K/c.$$

Therefore, we have converted the PDE into two ODEs. Solving, we get

$$T(t) = c_1e^{Kt}, \quad X(x) = c_2e^{-Kx/c},$$

so $u(x,t) = Ae^{Kt-Kx/c} = Ae^{-\frac{K}{c}(x-ct)}$, for some constants K and A (where A is shorthand for c_1c_2; in terms of D'Alembert's solution, $f(y) = Ae^{-\frac{K}{c}(y)}$). The general solution is a sum of these (for various A's and K's).

This can also be done in Sage:

Sage

```
sage: t = var("t")
sage: T = function("T",t)
sage: K = var("K")
sage: T0 = var("T0")
sage: sage: desolve(diff(T,t) ==  K*T, [T,t], [0,T0])
T0*e^(K*t)
sage: x = var("x")
sage: X = function("X",x)
sage: c = var("c")
sage: X0 = var("X0")
```

```
sage: desolve(diff(X,x) == -c^(-1)*K*X, [X,x], [0,X0])
X0*e^(-K*x/c)
sage: solnX =desolve(diff(X,x) == -c^(-1)*K*X, [X,x], [0,X0])
sage: solnX
X0*e^(-K*x/c)
sage: solnT = desolve(diff(T,t) ==  K*T, [T,t], [0,T0])
sage: solnT
T0*e^(K*t)
sage: solnT*solnX
T0*X0*e^(K*t - K*x/c)
```

Example 4.1.1. Assume water is flowing along a horizontal pipe at 3 gal/min in the x direction and that there is an initial concentration of salt distributed in the water with concentration of $u(x,0) = e^{-x}$. Using separation of variables, find the concentration at time t. Plot this for various values of t.

Solution. The method of separation of variables gives the "separated form" of the solution to the transport PDE as $u(x,t) = Ae^{Kt-Kx/c}$, where $c = 3$. The initial condition implies

$$e^{-x} = u(x,0) = Ae^{K\cdot 0-Kx/c} = Ae^{-Kx/3},$$

so $A = 1$ and $K = 3$. Therefore, $u(x,t) = e^{3t-x}$. In other words, the salt concentration is increasing in time. This makes sense if you think about it this way: "Freeze" the water motion at time $t = 0$. There is a lot of salt at the beginning of the pipe and less and less salt as you move along the pipe. Go down the pipe in the x-direction some distance to where you can barely tell there is any salt in the water. Now "unfreeze" the water motion. Looking along the pipe, you see the concentration increasing since the saltier water is now moving toward you.

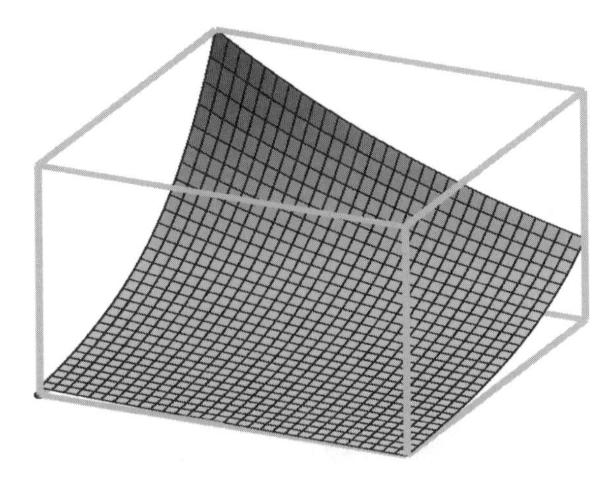

Figure 4.1: Transport with velocity $c = 3$.

An interactive version of the plot in Figure 4.1 can be produced in Sage with the following commands.

Sage

```
sage: t,x = var("t,x")
sage: plot3d(exp(3*t-x),[x,0,2],[t,0,2])
```

What if the initial concentration was not $u(x,0) = e^{-x}$ but instead $u(x,0) = e^{-x} + 3e^{-5x}$? How does the solution to

$$u_t + 3u_x = 0, \quad u(x,0) = e^{-x} + 3e^{-5x} \tag{4.1}$$

differ from the method of solution used above? In this case, we must use the fact that (by superposition) "the general solution" is of the form

$$u(x,t) = A_1 e^{K_1(t-x/3)} + A_2 e^{K_2(t-x/3)} + A_3 e^{K_3(t-x/3)} + \dots , \tag{4.2}$$

for some constants $A_1, K_1, \dots$. To solve this PDE (4.1), we must answer the following questions: (1) How many terms from (4.2) are needed? (2) What are the constants $A_1, K_1, \dots$? There are two terms in $u(x,0)$, so we can hope that we only need to use two terms and solve

$$e^{-x} + 3e^{-5x} = u(x,0) = A_1 e^{K_1(0-x/3)} + A_2 e^{K_2(0-x/3)}$$

for A_1, K_1, A_2, K_2. Indeed, this can be solved: $A_1 = 1$, $K_1 = 3$, $A_2 = 3$, $K_1 = 15$. This gives

$$u(x,t) = e^{3(t-x/3)} + 3e^{15(t-x/3)}.$$

4.1.2 The heat equation

We turn to the heat equation next,

$$k\frac{\partial^2 u(x,t)}{\partial x^2} = \frac{\partial u(x,t)}{\partial t}.$$

Example 4.1.2. First, assume the solution to the heat equation has the "factored" form

$$u(x,t) = X(x)T(t)$$

for some (unknown) functions X, T. If this function solves the PDE then it must satisfy $kX''(x)T(t) = X(x)T'(t)$, or

$$\frac{X''(x)}{X(x)} = \frac{1}{k}\frac{T'(t)}{T(t)}.$$

Since x and t are independent variables, these quotients must be constant. Think of it this way: the derivative of $\frac{X''(x)}{X(x)}$ *with respect to* t is zero. Therefore, the derivative of $\frac{1}{k}\frac{T'(t)}{T(t)}$ with respect to t is zero. This implies it is a constant. In other words, there must be a constant C such that

$$\frac{T'(t)}{T(t)} = kC, \quad X''(x) - CX(x) = 0.$$

Now we have reduced the problem of solving the one PDE to two ordinary differential equations (which is good), but with the price that we have introduced a constant that we don't know, namely, C (which maybe isn't so good). The first ordinary differential equation is easy to solve:

$$T(t) = A_1 e^{kCt}$$

for some constant A_1. It is best to analyze three cases now:

- *Case* $C = 0$: This implies $X(x) = A_2 + A_3 x$, for some constants A_2, A_3. Therefore

$$u(x,t) = A_1(A_2 + A_3 x) = \frac{a_0}{2} + b_0 x,$$

where (for reasons explained later) $A_1 A_2$ has been renamed $\frac{a_0}{2}$ and $A_1 A_3$ has been renamed b_0.

- *Case* $C < 0$: Write (for convenience) $C = -r^2$, for some $r > 0$. The ordinary differential equation for X implies $X(x) = A_2\cos(rx) + A_3\sin(rx)$, for some constants A_2, A_3. Therefore

$$u(x,t) = A_1 e^{-kr^2 t}(A_2\cos(rx) + A_3\sin(rx)) = (a\cos(rx) + b\sin(rx))e^{-kr^2 t},$$

where $A_1 A_2$ has been renamed a and $A_1 A_3$ has been renamed b.

- *Case* $C > 0$: Write (for convenience) $C = r^2$, for some r. The ordinary differential equation for X implies $X(x) = A_2 e^{rx} + A_3 e^{-(rx}$, for some constants A_2, A_3. Therefore

$$u(x,t) = e^{-kr^2 t}(ae^{rx} + be^{-rx}),$$

where $A_1 A_2$ has been renamed a and $A_1 A_3$ has been renamed b.

These are the solutions of the heat equation which can be written in factored form.

Exercises

1. Use `Sage` to solve and plot the solution to the following problem. Assume water is flowing along a horizontal pipe at 3 gal/min in the x-direction and that there is an initial concentration of salt distributed in the water with concentration of $u(x,0) = e^x$.

2. Use separation of variables to find all solutions to $u_t = \alpha u_x + u$, in factored form (where α is a nonzero constant).

3. Use separation of variables to find all solutions to $u_t + xu_x = u$, in factored form.

4. Use separation of variables to find all solutions to $u_t + tu_x = u$, in factored form.

5. Use separation of variables to find the unique solution to $u_t + 2u_x = 3u$, $u(x,0) = 7e^{-5x}$.

4.2 The method of superposition

Roughly speaking, the superposition principle says that the sum of two solutions to a linear differential equation is another solution. We give a specific example, in the following statement.

Theorem 4.2.1. *("Superposition")* If the solution to

$$\begin{cases} u_t = ku_{xx}, \\ u(x,0) = f_1(x) \end{cases}$$

is $u_1(x,t)$, and the solution to

$$\begin{cases} u_t = ku_{xx}, \\ u(x,0) = f_2(x) \end{cases}$$

is $u_2(x,t)$, then the solution to

$$\begin{cases} u_t &= ku_{xx}, \\ u(x,0) &= f_1(x) + f_2(x) \end{cases}$$

is $u_1(x,t) + u_2(x,t)$,

We consider a few specific instances of this, but first we need a few more facts.

For the "zero ends problem":

Theorem 4.2.2. If $n > 0$ is an integer and b_n is a given real number then the solution to

$$\begin{cases} u_t &= ku_{xx}, \\ u(x,0) &= b_n \sin(\frac{n\pi x}{L}), \\ u(0,t) &= u(L,t) = 0 \end{cases}$$

is $u(x,t) = b_n \sin(\frac{n\pi x}{L})e^{-k(\frac{n\pi}{L})^2 t}$.

This is not hard to verify directly: we must check that $b_n \sin(\frac{n\pi x}{L})e^{-k(\frac{n\pi}{L})^2 t}$ satisfies

1. the PDE $u_t = ku_{xx}$: $u_t = -k(\frac{n\pi}{L})^2 \cdot b_n \sin(\frac{n\pi x}{L})e^{-k(\frac{n\pi}{L})^2 t}$, $ku_{xx} = kb_n(-(\frac{n\pi}{L})^2 \sin(\frac{n\pi x}{L}))e^{-k(\frac{n\pi}{L})^2 t}$;

2. the initial condition: $u(x,0) = b_n \sin(\frac{n\pi x}{L})$ (take $t = 0$ and note $e^0 = 1$);

3. the boundary conditions: $u(0,t) = u(L,t) = 0$ (take $x = 0$ and note that $\sin(0) = 0$ or take $x = L$ and note that $\sin(n\pi) = 0$).

Example 4.2.1. The solution to

$$\begin{cases} \frac{1}{2}u_t &= u_{xx}, \\ u(x,0) &= -3\sin(11x), \\ u(0,t) &= u(\pi,t) = 0 \end{cases}$$

is $u(x,t) = -3\sin(11x)e^{-2\cdot 11^2 t} = -3\sin(11x)e^{-242t}$. The solution to

$$\begin{cases} \frac{1}{2}u_t &= u_{xx}, \\ u(x,0) &= 4\sin(5x), \\ u(0,t) &= u(\pi,t) = 0 \end{cases}$$

is $u(x,t) = 4\sin(5x)e^{-2\cdot 5^2 t} = 4\sin(5x)e^{-50t}$. Therefore, by superposition (Theorem 4.2.1), the solution to

$$\begin{cases} \frac{1}{2}u_t = u_{xx}, \\ u(x,0) = 4\sin(5x) - 3\sin(11x), \\ u(0,t) = u(\pi,t) = 0 \end{cases}$$

is $u(x,t) = 4\sin(5x)e^{-50t} - 3\sin(11x)e^{-242t}$.

For the "insulated ends problem", the analog of Theorem 4.2.2 is:

Theorem 4.2.3. The solution to

$$\begin{cases} u_t &= ku_{xx}, \\ u(x,0) &= a_n\cos(\frac{n\pi x}{L}), \\ u_x(0,t) &= u_x(L,t) = 0 \end{cases}$$

is $u(x,t) = a_n\cos(\frac{n\pi x}{L})e^{-k(\frac{n\pi}{L})^2 t}$.

Example 4.2.2. The solution to

$$\begin{cases} \frac{1}{2}u_t &= u_{xx}, \\ u(x,0) &= -3\cos(11x), \\ u_x(0,t) &= u_x(\pi,t) = 0 \end{cases}$$

is $u(x,t) = -3\cos(11x)e^{-2\cdot 11^2 t}$. The solution to

$$\begin{cases} \frac{1}{2}u_t &= u_{xx}, \\ u(x,0) &= 5, \\ u_x(0,t) &= u_x(\pi,t) = 0 \end{cases}$$

is $u(x,t) = 5$. Therefore, by superposition (Theorem 4.2.1), the solution to

$$\begin{cases} \frac{1}{2}u_t &= u_{xx}, \\ u(x,0) &= 5 - 3\cos(11x), \\ u_x(0,t) &= u_x(\pi,t) = 0 \end{cases}$$

is $u(x,t) = 5 - 3\cos(11x)e^{-2\cdot 11^2 t} = 5 - 3\cos(11x)e^{-242t}$.

Of course, there is no need to stop with only two terms. Using superposition, we know that the solution to the zero ends heat problem

$$\begin{cases} u_t = ku_{xx}, \\ u(x,0) = \sum_{n=1}^{\infty} b_n \sin(\frac{n\pi x}{L}), \\ u(0,t) = u(L,t) = 0 \end{cases} \tag{4.3}$$

is $u(x,t) = \sum_{n=1}^{\infty} b_n \sin(\frac{n\pi x}{L})e^{-k(\frac{n\pi}{L})^2 t}$. Likewise, the solution to the insulated ends heat problem

$$\begin{cases} u_t = ku_{xx}, \\ u(x,0) = \frac{a_0}{2} + \sum_{n=1}^{\infty} a_n \cos(\frac{n\pi x}{L}), \\ u_x(0,t) = u_x(L,t) = 0 \end{cases} \tag{4.4}$$

is $u(x,t) = \frac{a_0}{2} + \sum_{n=1}^{\infty} a_n \cos(\frac{n\pi x}{L})e^{-k(\frac{n\pi}{L})^2 t}$.

This last expression motivates the study of sine and cosine series studied in the next section.

Exercises

1. Verify Theorem 4.2.3.

2. Use superposition and Theorem 4.2.2 to solve

$$\begin{cases} u_t &= 3u_{xx}, \\ u(x,0) &= -\frac{1}{2}\sin(\frac{2\pi x}{5}) + 2\sin(\frac{3\pi x}{5}), \\ u(0,t) &= u(5,t) = 0. \end{cases}$$

3. Solve

$$\begin{cases} u_t &= 3u_{xx}, \\ u(x,0) &= \sum_{n=1}^{\infty} \frac{1}{n^2}\sin(\frac{n\pi x}{5}), \\ u(0,t) &= u(5,t) = 0. \end{cases}$$

4. Use superposition and Theorem 4.2.3 to solve

$$\begin{cases} u_t &= 3u_{xx}, \\ u(x,0) &= -\frac{1}{2}\cos(\frac{2\pi x}{5}) + 2\cos(\frac{3\pi x}{5}), \\ u_x(0,t) &= u_x(5,t) = 0. \end{cases}$$

5. Solve

$$\begin{cases} u_t &= 3u_{xx}, \\ u(x,0) &= 2 + \sum_{n=1}^{\infty} \frac{1}{n^2}\cos(\frac{n\pi x}{5}), \\ u_x(0,t) &= u_x(5,t) = 0. \end{cases}$$

4.3 Fourier, sine, and cosine series

This section introduces Fourier series, background for understanding Fourier's solution to the heat equation presented in §4.4.

4.3.1 Brief history

Fourier series were introduced by Joseph Fourier, a Frenchman who was a physicist, among other things. In fact, Fourier was Napoleon's scientific advisor during France's invasion of Egypt in the late 1800s. When Napoleon returned to France, he "elected" (i.e., appointed) Fourier to be a prefect—basically an important administrative post where he oversaw some large construction projects, such as highway development. It was during this time that Fourier worked on the theory of heat on the side. It turns out that D'Lambert looked at the wave equation some years earlier, using very similar techniques, but the approach was rejected (possibly due to its lack of rigor). However, Fourier's solution, explained in his 1807 memoir *On the Propagation of Heat in Solid Bodies*, was awarded a mathematics prize in 1811 by the Paris Institute. The gap in time is some indication of the controversy it caused at the time.

Fourier's solution to the heat equation is basically what undergraduates often learn in a differential equations class. The exception being that our understanding of Fourier series now is much better than what was known in the early 1800s and some of these facts, such as Dirichlet's theorem (see Theorem 4.3.1 below), are covered as well.

4.3.2 Motivation

Fourier series, sine series, and cosine series are all expansions for a given function $f(x)$ in terms of sines and/or cosines, somewhat analogous to the way that a Taylor series $a_0 + a_1x + a_2x^2 + \ldots$ is an expansion for a given function in powers of x. Both Fourier and Taylor series can be used to approximate the function they represent. However, there are at least three important differences between the two types of series.

1. For a function to have a Taylor series it must be differentiable,[2] whereas for a Fourier series it does not even have to be continuous.

2. Another difference is that the Taylor series is typically not periodic (though it can be in some cases), whereas a Fourier series is *always* periodic.

[2]Remember the formula for the nth Taylor series coefficient centered at $x = 0$: $a_n = f^{(n)}(0)/n!$.

3. Finally, the Taylor series (when it converges) always converges to the function $f(x)$, but the Fourier series may not (see Dirichlet's theorem below for a more precise description of what happens).

4.3.3 Definitions

Suppose $L > 0$ is given. A *Fourier series* of period $2L$ is a series of the form

$$\frac{a_0}{2} + \sum_{n=1}^{\infty} \left[a_n \cos\left(\frac{n\pi x}{L}\right) + b_n \sin\left(\frac{n\pi x}{L}\right) \right],$$

where the a_n and the b_n are given numbers. We shall implicitly assume that, for each x, this sum converges to some value, denoted $F(x)$. The *Nth partial sum* of this Fourier series is denoted

$$F_N(x) = \frac{a_0}{2} + \sum_{n=1}^{N} \left[a_n \cos\left(\frac{n\pi x}{L}\right) + b_n \sin\left(\frac{n\pi x}{L}\right) \right],$$

for $N > 0$. For the remainder of this chapter, we assume the a_n's and b_n's have the property that, for each x, $\lim_{N\to\infty} F_N(x) = F(x)$.

As a special case, if all the $a_n = 0$, then we call this Fourier series a *(Fourier) sine series.* Likewise, if all the $b_n = 0$, then we call this Fourier series a *(Fourier) cosine series.*

Fourier series

Let $f(x)$ be a function defined on an interval of the real line. We allow $f(x)$ to be discontinuous but the points in this interval where $f(x)$ is discontinuous must be finite in number and must be jump discontinuities.

First, we discuss Fourier series. To have a Fourier series you must be given two things:

- a "period" $P = 2L$;
- a function $f(x)$ defined on an interval of length $2L$, say $-L < x < L$ (but sometimes $0 < x < 2L$ is used instead).

The *Fourier series expansion of $f(x)$ with period $2L$* is

$$f(x) \sim \frac{a_0}{2} + \sum_{n=1}^{\infty} [a_n \cos(\frac{n\pi x}{L}) + b_n \sin(\frac{n\pi x}{L})], \tag{4.5}$$

where a_n and b_n are given by the formulas,[3]

$$a_n = \frac{1}{L} \int_{-L}^{L} f(x) \cos\left(\frac{n\pi x}{L}\right) dx, \tag{4.6}$$

[3]These formulas were not known to Fourier. To compute the Fourier coefficients a_n, b_n he used sometimes ingenious round-about methods involving large systems of equations.

and

$$b_n = \frac{1}{L}\int_{-L}^{L} f(x)\sin\left(\frac{n\pi x}{L}\right)\,dx. \tag{4.7}$$

Convergence and Dirichlet's theorem

The symbol $\sim$ is used in (4.5) above instead of $=$ because of the fact that the Fourier series may not converge to $f(x)$ (see Dirichlet's theorem below). Do you remember right-hand and left-hand limits from Calculus I? Recall that they are denoted $f(x+) = \lim_{\epsilon\to 0,\epsilon>0} f(x+\epsilon)$ and $f(x-) = \lim_{\epsilon\to 0,\epsilon>0} f(x-\epsilon)$, respectively. We know now that the Fourier series does not converge to the value of $f(x)$ at every point where x is a point of jump discontinuity.[4] The convergence properties are given by the theorem below.

Theorem 4.3.1. (*Dirichlet's[5] theorem*) Let $f(x)$ be a function defined on $-L < x < L$ which is bounded and piecewise continuous with only finitely many discontinuities. The Fourier series of $f(x)$,

$$f(x) \sim \frac{a_0}{2} + \sum_{n=1}^{\infty}\left[a_n\cos\left(\frac{n\pi x}{L}\right) + b_n\sin\left(\frac{n\pi x}{L}\right)\right]$$

(where a_n and b_n are as in the formulas (4.6) and (4.7)) converges to

$$\frac{f(x+) + f(x-)}{2}.$$

In other words, the Fourier series of $f(x)$ converges to $f(x)$ only if $f(x)$ is continuous at x. If $f(x)$ is not continuous at x then the Fourier series of $f(x)$ converges to the "midpoint of the jump."

Of course, since a sine series or cosine series is a special case of a Fourier series, Dirichlet's theorem applies to them as well.

Example 4.3.1. Find the Fourier series of $f(x) = 2 + x$, $-2 < x < 2$.

The definition of L implies $L = 2$. Without even computing the Fourier series, we can evaluate it using Dirichlet's theorem.

Exercise

1. Using periodicity and Dirichlet's theorem, find the value to which the Fourier series of $f(x)$ converges at $x = 1, 2, 3$.

Solution. $f(x)$ is continuous at 1, so the Fourier series at $x = 1$ converges to $f(1) = 3$ by Dirichlet's theorem. $f(x)$ is not defined at 2. Its Fourier series is periodic with period 4, so at $x = 2$ the Fourier series converges to $\frac{f(2+)+f(2-)}{2} = \frac{0+4}{2} = 2$. $f(x)$ is not defined at

[4] Fourier believed "his" series converged to the function in the early 1800s but we now know this is not always true.

[5] Pronounced "Dear-ish-lay".

3. Its Fourier series is periodic with period 4, so at $x = 3$ the Fourier series converges to $\frac{f(-1)+f(-1+)}{2} = \frac{1+1}{2} = 1$. This solves the problem.

The formulas (4.6) and (4.7) enable us to compute the Fourier series coefficients a_0, a_n and b_n. (We skip the details.) These formulas give that the Fourier series of $f(x)$ is

$$f(x) \sim \frac{4}{2} + \sum_{n=1}^{\infty} -4\,\frac{n\pi\cos(n\pi)}{n^2\pi^2}\sin\left(\frac{n\pi x}{2}\right) = 2 - \frac{4}{\pi}\sum_{n=1}^{\infty}\frac{\cos(n\pi)}{n}\sin\left(\frac{n\pi x}{2}\right).$$

The Fourier series approximations to $f(x)$ are

$$S_0 = 2,\ S_1 = 2 + \frac{4}{\pi}\sin\left(\frac{\pi x}{2}\right),\ S_2 = 2 + 4\,\frac{\sin\left(\frac{1}{2}\pi x\right)}{\pi} - 2\,\frac{\sin(\pi x)}{\pi},\ \ldots$$

The graphs of each of these functions get closer and closer to the graph of $f(x)$ on the interval $-2 < x < 2$. For instance, the graphs of $f(x)$ and of S_8 are given in Figure 4.2.

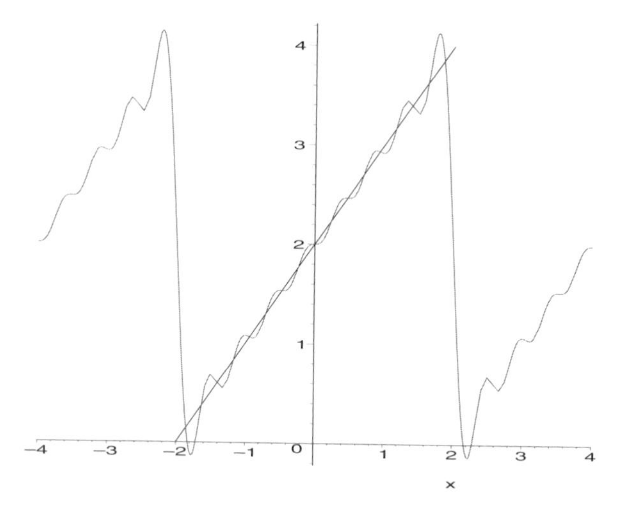

Figure 4.2: Graph of $f(x)$ and a Fourier series approximation of $f(x)$.

Notice that $f(x)$ is defined only from $-2 < x < 2$ yet the Fourier series not only is defined everywhere but is periodic with period $P = 2L = 4$. Also, notice that S_8 is not a bad approximation to $f(x)$.

This can also be done in Sage. First, we define the function.

Sage

```
sage: f = lambda x:x+2
sage: f = Piecewise([[(-2,2),f]])
```

This can be plotted using the command `f.plot().show()`. Next, we compute the Fourier series coefficients:

Sage

```
sage: f.fourier_series_cosine_coefficient(0,2) # a_0
4
sage: f.fourier_series_cosine_coefficient(1,2) # a_1
0
sage: f.fourier_series_cosine_coefficient(2,2) # a_2
0
sage: f.fourier_series_cosine_coefficient(3,)  # a_3
0
sage: f.fourier_series_sine_coefficient(1,2) # b_1
4/pi
sage: f.fourier_series_sine_coefficient(2,) # b_2
-2/pi
sage: f.fourier_series_sine_coefficient(3,2) # b_3
4/(3*pi)
```

Finally, the partial Fourier series and its plot versus the function can be computed using the following Sage commands.

Sage

```
sage: f.fourier_series_partial_sum(3,2)
-2*sin(pi*x)/pi + 4*sin(pi*x/2)/pi + 2
sage: P1 = f.plot_fourier_series_partial_sum(15,2,-5,5,linestyle=":")
sage: P2 = f.plot(rgbcolor=(1,1/4,1/2))
sage: (P1+P2).show()
```

The plot (which takes 15 terms of the Fourier series) is given in Figure 4.2.

Cosine series

Next, we discuss cosine series. To have a cosine series you must be given two things:

- a period $P = 2L$, and
- a function $f(x)$ defined on the interval of length L, $0 < x < L$.

The *cosine series expansion of $f(x)$ with period $2L$* is

$$f(x) \sim \frac{a_0}{2} + \sum_{n=1}^{\infty} a_n \cos(\frac{n\pi x}{L}),$$

where a_n is given by

$$a_n = \frac{2}{L} \int_0^L \cos(\frac{n\pi x}{L}) f(x)\, dx. \tag{4.8}$$

The cosine series of $f(x)$ is exactly the same as the Fourier series of the *even extension* of $f(x)$, defined by

$$f_{even}(x) = \begin{cases} f(x), & 0 < x < L, \\ f(-x), & -L < x < 0. \end{cases}$$

Example 4.3.2. Let's consider an example of a cosine series. In this case, we take the piecewise-constant function $f(x)$ defined on $0 < x < 3$ by

$$f(x) = \begin{cases} 1, & 0 < x < 2, \\ -1, & 2 \le x < 3. \end{cases}$$

We see therefore that $L = 3$. The formula above for the cosine series coefficients gives the result that

$$f(x) \sim \frac{1}{3} + \sum_{n=1}^{\infty} 4\,\frac{\sin\left(\frac{2}{3}n\pi\right)}{n\pi}\cos\left(\frac{n\pi x}{3}\right).$$

The first few partial sums are

$$S_2 = 1/3 + 2\,\frac{\sqrt{3}\cos\left(\frac{1}{3}\pi x\right)}{\pi},$$

$$S_3 = 1/3 + 2\,\frac{\sqrt{3}\cos\left(\frac{1}{3}\pi x\right)}{\pi} - \frac{\sqrt{3}\cos\left(\frac{2}{3}\pi x\right)}{\pi},$$

$$\vdots$$

As before, the more terms in the cosine series we take, the better is the approximation, for $0 < x < 3$. Comparing Figure 4.3 with Figure 4.2, note that even with more terms, this approximation is not as good as the previous example. The precise reason for this is rather technical but basically boils down to the following: roughly speaking, the more differentiable the function is, the faster the Fourier series converges (and therefore the better the partial sums of the Fourier series will approximate $f(x)$). Also, notice that the cosine series approximation S_{10} is an even function but $f(x)$ is not (its defined only in $0 < x < 3$).

For instance, the graphs of $f(x)$ and of S_{10} are given in Figure 4.3.

Sine series

Finally, we define sine series. As with a cosine series, to have a sine series you must be given two things:

- a period $P = 2L$, and
- a function $f(x)$ defined on the interval of length L, $0 < x < L$.

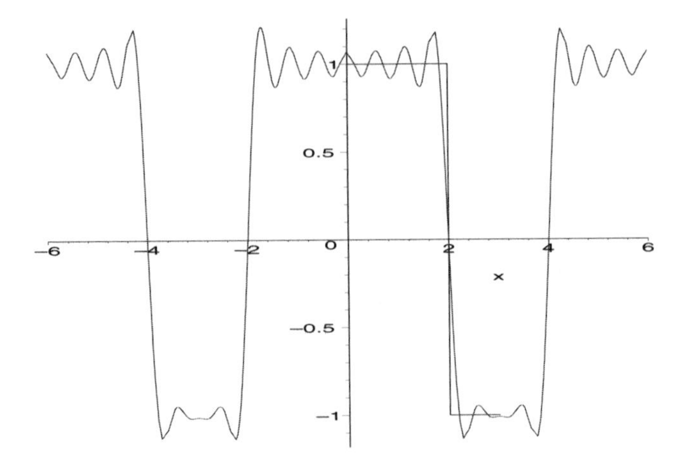

Figure 4.3: Graph of $f(x)$ and a cosine series approximation of $f(x)$.

The *sine series of* $f(x)$ *with period* $2L$ is

$$f(x) \sim \sum_{n=1}^{\infty} b_n \sin\left(\frac{n\pi x}{L}\right),$$

where b_n is given by

$$b_n = \frac{2}{L}\int_0^L \sin\left(\frac{n\pi x}{L}\right) f(x)\, dx. \tag{4.9}$$

The sine series of $f(x)$ is exactly the same as the Fourier series of the *odd extension* of $f(x)$, defined by

$$f_{odd}(x) = \begin{cases} f(x), & 0 < x < L, \\ -f(-x), & -L < x < 0. \end{cases}$$

Example 4.3.3. Let's consider an example of a sine series. In this case, we take the piecewise-constant function $f(x)$ defined on $0 < x < 3$ by the same expression we used in the cosine series example above.

Exercise

1. Using periodicity and Dirichlet's theorem, find the value that the sine series (SS) of $f(x)$ converges to at $x = 1, 2, 3$.

Solution. $f(x)$ is continuous at 1, so the Fourier series at $x = 1$ converges to $f(1) = 1$. $f(x)$ is not continuous at 2, so at $x = 2$ the SS converges to $\frac{f(2+)+f(2-)}{2} = \frac{f(-2+)+f(2-)}{2} = \frac{-1+1}{2} = 0$. $f(x)$ is not defined at 3. Its SS is periodic with period 6, so at $x = 3$ the SS converges to $\frac{f_{odd}(3-)+f_{odd}(3+)}{2} = \frac{-1+1}{2} = 0$. This solves the problem.

The formula above for the sine series coefficients gives the result

$$f(x) = \sum_{n=1}^{\infty} 2\,\frac{\cos(n\pi) - 2\cos\left(\frac{2}{3}n\pi\right) + 1}{n\pi} \sin\left(\frac{n\pi x}{3}\right).$$

The partial sums are

$$S_2 = 2\,\frac{\sin(1/3\,\pi\,x)}{\pi} + 3\,\frac{\sin\left(\frac{2}{3}\,\pi\,x\right)}{\pi},$$

$$S_3 = 2\,\frac{\sin\left(\frac{1}{3}\,\pi\,x\right)}{\pi} + 3\,\frac{\sin\left(\frac{2}{3}\,\pi\,x\right)}{\pi} - 4/3\,\frac{\sin(\pi\,x)}{\pi},$$

$$\vdots$$

These partial sums S_n, as $n \to \infty$, converge to their limit about as fast as those in the previous example. Instead of taking only ten terms, this time we take forty. Observe from the graph in Figure 4.4 that the value of the sine series at $x = 2$ does seem to be approaching 0, as Dirichlet's theorem predicts. The graph of $f(x)$ with S_{40} is given in Figure 4.4.

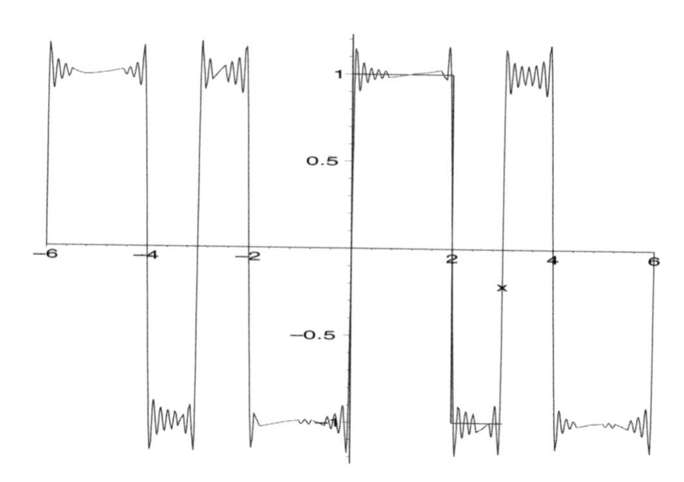

Figure 4.4: Graph of $f(x)$ and a sine series approximation of $f(x)$.

Exercises

1. Let $f(x) = x^2$, $-2 < x < 2$ and $L = 2$. Use Sage to compute the first ten terms of the Fourier series, and plot the corresponding partial sum. Next plot the partial sum of the first fifty terms and compare them.

2. What mathematical results do the following Sage commands give you? In other words, if you see someone typing these commands into a computer, what problem they are trying to solve?

Sage

```
sage: x = var("x")
sage: f0(x) = 0
sage: f1(x) = -1
sage: f2(x) = 1
sage: f = Piecewise([[(-2,0),f1],[(0,3/2),f0],[(3/2,2),f2]])
sage: P1 = f.plot()
sage: a10 = [f.fourier_series_cosine_coefficient(n,2) for n in range(10)]
sage: b10 = [f.fourier_series_sine_coefficient(n,2) for n in range(10)]
sage: fs10 = a10[0]/2 + sum([a10[i]*cos(i*pi*x/2) for i in range(1,10)])
 + sum([b10[i]*sin(i*pi*x/2) for i in range(10)])
sage: P2 = fs10.plot(-4,4,linestyle=":")
sage: (P1+P2).show()
sage: ### these commands below are more time-consuming:
sage: a50 = [f.fourier_series_cosine_coefficient(n,2) for n in range(50)]
sage: b50 = [f.fourier_series_sine_coefficient(n,2) for n in range(50)]
sage: fs50 = a50[0]/2 + sum([a50[i]*cos(i*pi*x/2) for i in range(1,50)])
 + sum([b50[i]*sin(i*pi*x/2) for i in range(50)])
sage: P3 = fs50.plot(-4,4,linestyle="--")
sage: (P1+P2+P3).show()
sage: a100 = [f.fourier_series_cosine_coefficient(n,2) for n in range(100)]
sage: b100 = [f.fourier_series_sine_coefficient(n,2) for n in range(100)]
sage: fs100 = a100[0]/2 + sum([a100[i]*cos(i*pi*x/2) for i in range(1,100)])
 + sum([b100[i]*sin(i*pi*x/2) for i in range(100)])
sage: P3 = fs100.plot(-4,4,linestyle="--")
sage: (P1+P2+P3).show()
```

3. A periodic function is given over one period by $f(t) = |t|, -\pi < t < \pi$. Which of the following statements is correct for the Fourier series $\frac{a_0}{2} + \sum_{n=1}^{\infty} a_n \cos\frac{n\pi t}{L} + b_n \sin\frac{n\pi t}{L}$ of $f(t)$?

   (a) $a_n = 0$ for all even integers n, but not for any odd integers n.

   (b) $a_n = 0$ for all odd integers n, but not for any even integers n.

   (c) $a_n = 0$ for $n = 0, 1, 2, \ldots$.

   (d) $a_n = 0$ for $n = 1, 2, 3, \ldots$, but $a_0 \neq 0$.

4. At the point $t = 0$ the Fourier sine series of the function $f(t) = 1, -0 \leq t \leq 1$ converges to what value? (a) -1, (b) 0, (c) $\frac{1}{2}$, (d) 1.

5. Let $f(x) = \begin{cases} 1 & \text{if } 0 < x < 1/2 \\ 0 & \text{if } 1/2 < x < 1. \end{cases}$

   (a) Find the Fourier sine series for $f(x)$.

   (b) To what value does the Fourier sine series converge at (i) $x = 1$? (ii) $x = \frac{3}{2}$? (iii) $x = \frac{1}{2}$?

6. Let $f(x) = x(\pi - x)$, $-\pi < x < \pi$.

   Find the cosine series of $f(x)$, $0 < x < \pi$, having period 2π, *and* graph the function to which the series converges for $-2\pi < x < 2\pi$.

7. Let

$$f(x) = \begin{cases} 1, & 0 \le t \le \pi/2, \\ -1, & \pi/2 < t < \pi, \end{cases}$$

period 2π.

   (a) Find the cosine series of $f(x)$ and graph it, $-2\pi < x < 2\pi$.

   (b) Write down the series in summation notation and write down the sum of the first three nonzero terms.

4.4 The heat equation

> The differential equations of the propagation of heat express the most general conditions, and reduce the physical questions to problems of pure analysis, and this is the proper object of theory.
>
> —*Jean-Baptist-Joseph Fourier*

The heat equation is the PDE $k\frac{\partial^2 u(x,t)}{\partial x^2} = \frac{\partial u(x,t)}{\partial t}$. This is a diffusion equation whose study is very important for many topics (even in subjects as diverse as financial mathematics, chemistry, and modeling criminal behavior!)

The heat equation with *zero ends* boundary conditions models the temperature of an (insulated) wire of length L:

$$\begin{cases} k\frac{\partial^2 u(x,t)}{\partial x^2}, & = \frac{\partial u(x,t)}{\partial t} \\ u(x,0) & = f(x). \\ u(0,t) & = u(L,t) = 0. \end{cases}$$

Here $u(x,t)$ denotes the temperature at a point x on the wire at time t. The initial temperature $f(x)$ is a given more or less arbitrary function[6] that is related to our temperature function $u(x,t)$ by the condition

$$u(x,0) = f(x).$$

In this model, it is assumed that no heat escapes out of the middle of the wire (which, say, is coated with some kind of insulating plastic). However, the boundary conditions $u(0,t) = u(L,t) = 0$ permit heat to dissipate or "escape" out of the ends.

[6]It must be "nice enough" that it has a Fourier series, so piecewise continuous will work for us.

4.4.1 Method for zero ends

The "recipe" for solving a zero ends heat problem by Fourier's method is pretty simple.

- Find the sine series of $f(x)$:

$$f(x) \sim \sum_{n=1}^{\infty} b_n(f) \sin\left(\frac{n\pi x}{L}\right),$$

  where the coefficients can be computed from (4.9).

- The solution is[7]

$$u(x,t) = \sum_{n=1}^{\infty} b_n(f) \sin\left(\frac{n\pi x}{L}\right) \exp\left(-k\left(\frac{n\pi}{L}\right)^2 t\right). \tag{4.10}$$

Example 4.4.1. Let

$$f(x) = \begin{cases} -1, & 0 \le x \le \pi/2, \\ 2, & \pi/2 < x < \pi. \end{cases}$$

Then $L = \pi$ and

$$b_n(f) = \frac{2}{\pi}\int_0^{\pi} f(x)\sin(nx)dx = -2\,\frac{2\,\cos(n\pi) - 3\,\cos(\frac{1}{2}\,n\pi) + 1}{n\pi}.$$

Thus

$$f(x) \sim b_1(f)\sin(x) + b_2(f)\sin(2x) + \cdots = \frac{2}{\pi}\sin(x) - \frac{6}{\pi}\sin(2x) + \frac{2}{3\pi}\sin(3x) + \dots.$$

This can also be done in Sage:

Sage

```
sage: x = var("x")
sage: f1(x) = -1
sage: f2(x) = 2
sage: f = Piecewise([[(0,pi/2),f1],[(pi/2,pi),f2]])
sage: P1 = f.plot()
sage: b10 = [f.sine_series_coefficient(n,pi) for n in range(1,10)]
sage: b10
[2/pi, -6/pi, 2/(3*pi), 0, 2/(5*pi), -2/pi, 2/(7*pi), 0, 2/(9*pi)]
sage: ss10 = sum([b10[n]*sin((n+1)*x) for n in range(len(b10))])
sage: ss10
2*sin(9*x)/(9*pi) + 2*sin(7*x)/(7*pi) - 2*sin(6*x)/pi
+ 2*sin(5*x)/(5*pi) + 2*sin(3*x)/(3*pi) - 6*sin(2*x)/pi + 2*sin(x)/pi
sage: b50 = [f.sine_series_coefficient(n,pi) for n in range(1,50)]
```

[7] A quick reminder: $\exp(x) = e^x$.

```
sage: ss50 = sum([b50[n]*sin((n+1)*x) for n in range(len(b50))])
sage: P2 = ss10.plot(-5,5,linestyle="--")
sage: P3 = ss50.plot(-5,5,linestyle=":")
sage: (P1+P2+P3).show()
```

This illustrates how the series converges to the function. The function $f(x)$, and some of the partial sums of its sine series, looks like Figure 4.5.

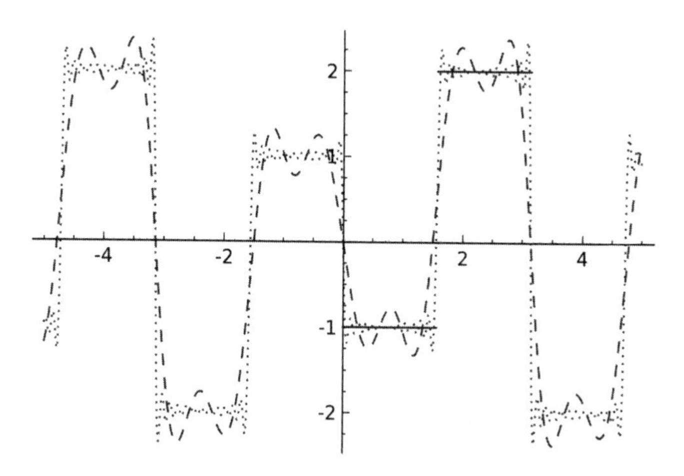

Figure 4.5: $f(x)$ and two sine series approximations.

As you can see, taking more and more terms gives partial sums which better and better approximate $f(x)$.

The solution to the heat equation, therefore, is (4.10).

Next, we see how Sage can plot the solution to the heat equation (we use $k = 1$):

Sage

```
sage: t = var("t")
sage: soln50(t) = sum([b50[n]*sin((n+1)*x)*e^(-(n+1)^2*t) for n in range(49)])
sage: P4 = plot(soln50(1/10) ,0,pi,linestyle=":")
sage: P5 = plot(soln50(1/2) ,0,pi)
sage: P6 = plot(soln50(1/10) ,0,pi,linestyle="--")
sage: (P1+P4+P5+P6).show()
```

Taking fifty terms of this series, the graph of the solution at $t = 0$, $t = 0.5$, $t = 1$, looks approximately like Figure 4.6.

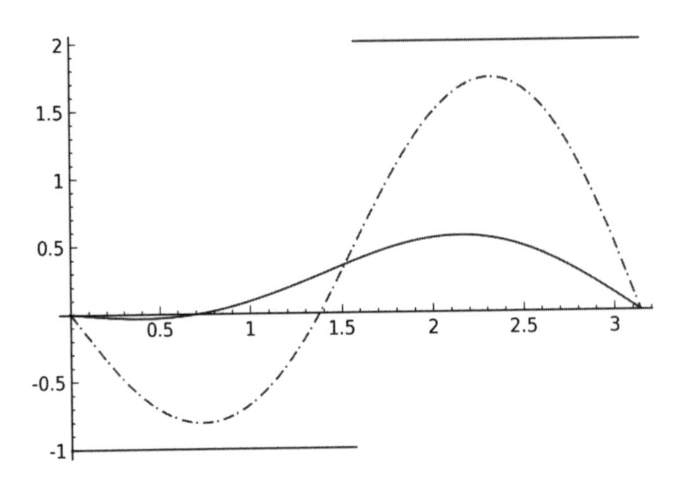

Figure 4.6: $f(x)$, $u(x,0.1)$, $u(x,0.5)$, $u(x,1.0)$ using sixty terms of the sine series.

4.4.2 Method for insulated ends

The heat equation with *insulated ends* boundary conditions models the temperature of an (insulated) wire of length L:

$$\begin{cases} k\frac{\partial^2 u(x,t)}{\partial x^2} &= \frac{\partial u(x,t)}{\partial t}, \\ u(x,0) &= f(x), \\ u_x(0,t) &= u_x(L,t) = 0. \end{cases}$$

Here $u_x(x,t)$ denotes the partial derivative of the temperature at a point x on the wire at time t. The initial temperature $f(x)$ is specified by the equation $u(x,0) = f(x)$.

- Find the cosine series of $f(x)$:

$$f(x) \sim \frac{a_0}{2} + \sum_{n=1}^{\infty} a_n(f) \cos\left(\frac{n\pi x}{L}\right),$$

  where the coefficients can be computed from (4.8).

- The solution is

$$u(x,t) = \frac{a_0}{2} + \sum_{n=1}^{\infty} a_n(f) \cos\left(\frac{n\pi x}{L}\right) \exp\left(-k\left(\frac{n\pi}{L}\right)^2 t\right). \tag{4.11}$$

Example 4.4.2. Let

$$f(x) = \begin{cases} -1, & 0 \leq x \leq \pi/2, \\ 2, & \pi/2 < x < \pi. \end{cases}$$

Then $L = \pi$ and

$$a_n(f) = \frac{2}{\pi}\int_0^\pi f(x)\cos(nx)dx = -6\frac{\sin\left(\frac{1}{2}\pi n\right)}{\pi n},$$

for $n > 0$ and $a_0 = 1$.

Thus

$$f(x) \sim \frac{a_0}{2} + a_1(f)\cos(x) + a_2(f)\cos(2x) + \dots$$

This can also be done in **Sage**:

Sage

```
sage: x = var("x")
sage: f1(x) = -1
sage: f2(x) = 2
sage: f = Piecewise([[(0,pi/2),f1],[(pi/2,pi),f2]])
sage: P1 = f.plot()
sage: a10 = [f.cosine_series_coefficient(n,pi) for n in range(10)]
sage: a10
[1, -6/pi, 0, 2/pi, 0, -6/(5*pi), 0, 6/(7*pi), 0, -2/(3*pi)]
sage: a50 = [f.cosine_series_coefficient(n,pi) for n in range(50)]
sage: cs10 = a10[0]/2 + sum([a10[n]*cos(n*x) for n in range(1,len(a10))])
sage: P2 = cs10.plot(-5,5,linestyle="--")
sage: cs50 = a50[0]/2 + sum([a50[n]*cos(n*x) for n in range(1,len(a50))])
sage: P3 = cs50.plot(-5,5,linestyle=":")
sage: (P1+P2+P3).show()
```

This illustrates how the series converges to the function. The piecewise-constant function $f(x)$, and some of the partial sums of its cosine series (one using ten terms and one using fifty terms), looks like Figure 4.7.

As you can see from Figure 4.7, taking more and more terms gives functions which better and better approximate $f(x)$.

Consider the heat problem

$$\begin{cases} \frac{\partial^2 u(x,t)}{\partial x^2} = \frac{\partial u(x,t)}{\partial t}, \\ u(x,0) = f(x), \\ u_x(0,t) = u_x(\pi,t) = 0. \end{cases}$$

The solution to the heat equation, therefore, is (4.11).

Using **Sage**, we can plot this function:

Sage

```
sage: soln50a = a50[0]/2 + sum([a50[n]*cos(n*x)*e^(-n^2*(1/100)) for n in range(1,len(a50))])
sage: soln50b = a50[0]/2 + sum([a50[n]*cos(n*x)*e^(-n^2*(1/10)) for n in range(1,len(a50))])
sage: soln50c = a50[0]/2 + sum([a50[n]*cos(n*x)*e^(-n^2*(1/2)) for n in range(1,len(a50))])
sage: P4 = soln50a.plot(0,pi)
```

```
sage: P5 = soln50b.plot(0,pi,linestyle=":")
sage: P6 = soln50c.plot(0,pi,linestyle="--")
sage: (P1+P4+P5+P6).show()
```

Taking only the first fifty terms of this series, the graph of the solution at $t = 0$, $t = 0.01$, $t = 0.1$,, $t = 0.5$, looks approximately like the graph in Figure 4.8.

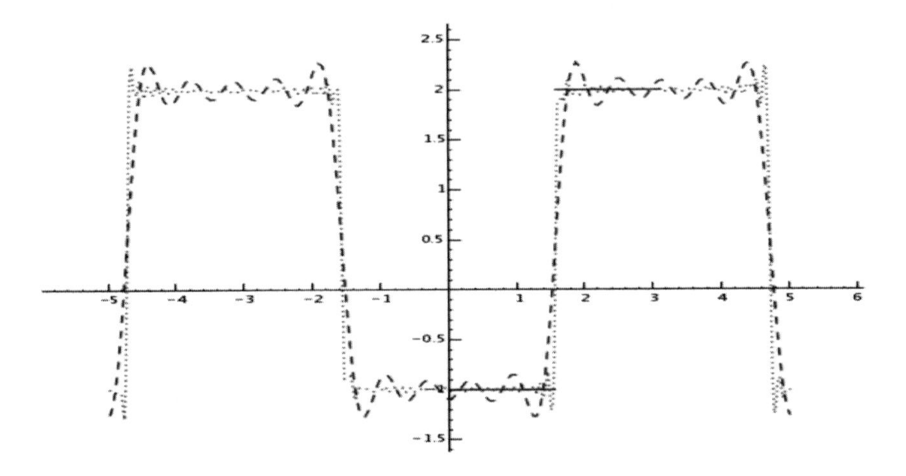

Figure 4.7: $f(x)$ and two cosine series approximations.

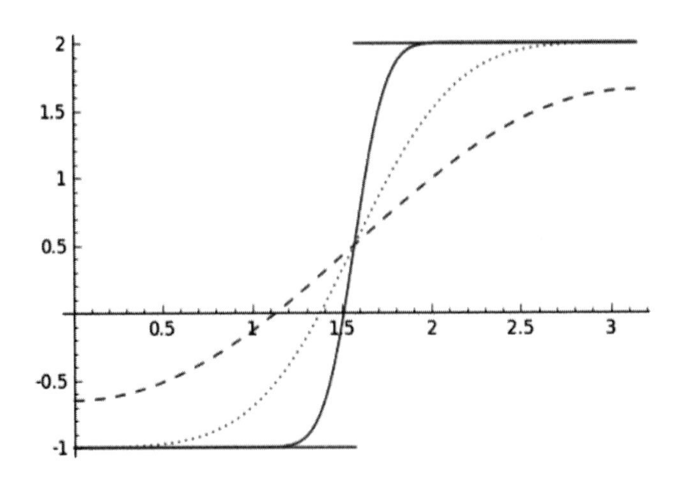

Figure 4.8: $f(x) = u(x, 0)$, $u(x, 0.01)$, $u(x, 0.1)$, $u(x, 0.5)$ using 50 terms of the cosine series.

4.4.3 Explanation via separation of variables

Where does this solution come from? It comes from the method of separation of variables and the superposition principle. Here is a short explanation. We shall discuss only the zero ends case (the insulated ends case is similar).

We saw in Example 4.1.2 that the partial differential equation

$$k\frac{\partial^2 u(x,t)}{\partial x^2} = \frac{\partial u(x,t)}{\partial t}$$

has solutions in "factored" form

$$u(x,t) = X(x)T(t),$$

where

$$T(t) = A_1 e^{kCt}$$

for some constant A_1 and $X(x)$ was described using a case-by-case analysis. Physical intuition tells us that $u(x,t)$ is bounded as $t \to \infty$. This implies $C \leq 0$, which rules out the third case in Example 4.1.2. By superposition, the general solution is a sum of these:

$$\begin{aligned} u(x,t) &= \tfrac{a_0}{2} + b_0 x + \textstyle\sum_{n=1}^{\infty}(a_n \cos(r_n x) + b_n \sin(r_n x))e^{-kr_n^2 t} \\ &= \tfrac{a_0}{2} + b_0 x + (a_1 \cos(r_1 x) + b_1 \sin(r_1 x))e^{-kr_1^2 t} \\ &\quad +(a_2 \cos(r_2 x) + b_2 \sin(r_2 x))e^{-kr_2^2 t} + \dots, \end{aligned} \tag{4.12}$$

for some a_i, b_i, r_i. We may order the r_i's to be strictly increasing if we like.

We have not yet used the initial condition $u(x,0) = f(x)$ or the boundary conditions $u(0,t) = u(L,t) = 0$. We do that next.

What do the boundary conditions tell us? Plugging $x = 0$ into (4.12) gives

$$0 = u(0,t) = \frac{a_0}{2} + \sum_{n=1}^{\infty} a_n e^{-kr_n^2 t} = \frac{a_0}{2} + a_1 e^{-kr_1^2 t} + a_2 e^{-kr_2^2 t} + \dots .$$

These exponential functions are linearly independent, so $a_0 = 0$, $a_1 = 0$, $a_2 = 0$, $\dots$. This implies

$$u(x,t) = b_0 x + \sum_{n=1} b_n \sin(r_n x)e^{-kr_n^2 t} = b_0 x + b_1 \sin(r_1 x)e^{-kr_1^2 t} + b_2 \sin(r_2 x)e^{-kr_2^2 t} + \dots .$$

Plugging $x = L$ into this gives

$$0 = u(L,t) = b_0 L + \sum_{n=1} b_n \sin(r_n L)e^{-kr_n^2 t}.$$

Again, exponential functions are linearly independent, so $b_0 = 0$, $b_n \sin(r_n L) = 0$ for $n = 1, 2, \dots$. In other to get a nontrivial solution to the PDE, we don't want $b_n = 0$, so $\sin(r_n L) = 0$. This forces $r_n L$ to be a multiple of π, say $r_n = n\pi/L$. This gives

$$u(x,t) = \sum_{n=1}^{\infty} b_n \sin\left(\frac{n\pi}{L}x\right) e^{-k(\frac{n\pi}{L})^2 t} = b_1 \sin\left(\frac{\pi}{L}x\right) e^{-k(\frac{\pi}{L})^2 t} + b_2 \sin\left(\frac{2\pi}{L}x\right) e^{-k(\frac{2\pi}{L})^2 t} + \dots, \tag{4.13}$$

for some b_i's. The special case $t = 0$ is the so-called sine series expansion of the initial temperature function $u(x,0)$. This was discovered by Fourier. To solve the heat eqution, it remains to solve for the sine series coefficients b_i.

There is one remaining condition which our solution $u(x,t)$ must satisfy.

What does the initial condition tell us? Plugging $t = 0$ into (4.13) gives

$$f(x) = u(x,0) = \sum_{n=1}^{\infty} b_n \sin(\frac{n\pi}{L}x) = b_1 \sin(\frac{\pi}{L}x) + b_2 \sin(\frac{2\pi}{L}x) + \dots .$$

In other words, if $f(x)$ is given as a sum of these sine functions, or if we can somehow express $f(x)$ as a sum of sine functions, then we can solve the heat equation. In fact there is a formula[8] for these coefficients b_n:

$$b_n = \frac{2}{L}\int_0^L f(x)\sin\left(\frac{n\pi}{L}x\right)dx.$$

It is this formula that is used in the solutions above.

Exercises

1. Solve the heat equation

$$\begin{cases} 2\frac{\partial^2 u(x,t)}{\partial x^2} = \frac{\partial u(x,t)}{\partial t}, \\ u_x(0,t) = u_x(3,t) = 0, \\ u(x,0) = x, \end{cases}$$

   using Sage to plot approximations as above.

2. A metal bar of length 2 and thermal diffusivity 3 has its left end held at 0° and its right end insulated for all times $t > 0$. Mathematically this is described by the partial differential equation and boundary conditions

$$\frac{\partial u}{\partial t} = 3\frac{\partial^2 u}{\partial x^2}, \quad u(0,t) = 0, \quad \frac{\partial u}{\partial x}(2,t) = 0.$$

   The physically reasonable solutions of the form $u(x,t) = X(x)T(t)$ that satisfy the partial differential equation are $u(x.t) = e^{-3\lambda^2 t}(A\cos\lambda x + B\sin\lambda x)$. The solutions of this form that also satisfy the boundary conditions are $u(x,t) =$

   (a) $e^{-3(\frac{n\pi}{2})^2 t} B \sin\frac{n\pi x}{2}, n = 1,2,3,\dots$; (b) $e^{-3(\frac{n\pi}{2})^2 t} A \cos\frac{n\pi x}{2}, n = 1,2,3,\dots$;
   (c) $e^{-3(\frac{n\pi}{4})^2 t} B \sin\frac{n\pi x}{4}, n = 1,3,5,\dots$; (d) $e^{-3(\frac{n\pi}{4})^2 t} A \cos\frac{n\pi x}{4}, n = 1,3,5,\dots$.

[8] Fourier did not know this formula at the time; it was discovered later by Dirichlet.

3. A thin bar of length 4 and thermal diffusivity $k = 1$, which initially has a temperature distribution given as a function of position by $5x$, has its ends insulated for times $t > 0$. This situation is described mathematically by

$$\frac{\partial u}{\partial t} = \frac{\partial^2 u}{\partial x^2},$$
$$\frac{\partial u}{\partial x}(0,t) = \frac{\partial u}{\partial x}(4,t) = 0, \ t > 0,$$
$$u(x,0) = 5x, \ 0 \le x \le 4.$$

- Find the temperature $u(x,t)$. Show all steps of the separation of variables process clearly.
- What is the final temperature distribution on the bar as $t \to \infty$?

4. Let $f(x) = \begin{cases} 1 & \text{if } 0 < x < 1/2, \\ 0 & \text{if } 1/2 < x < 1. \end{cases}$ Find the function $u(x,t)$ satisfying the heat equation

$$2\frac{\partial^2 u}{\partial x^2} = \frac{\partial u}{\partial t},$$
$$u(x,0) = f(x),$$
$$u(0,t) = u(1,t) = 0.$$

Write your answer in summation notation. Estimate $u(1/2,1)$ using the first three nonzero terms of the series solution.

5. Solve the heat equation

$$\frac{\partial u(x,t)}{\partial t} = \frac{1}{10}\frac{\partial^2 u(x,t)}{\partial x^2}$$

with insulated end points and $u(x,0) = x(\pi - x)$, $0 < x < \pi$. Show all steps of the separation of variables process.

6. The temperature distribution $u(x,t)$ on a thin bar satisfies the following conditions

$$u_{xx} = u_t, \quad u_x(0,t) = u_x(\pi,t) = 0, \quad u(x,0) = f(x),$$

where

$$f(x) = \begin{cases} 1, & 0 \le t \le \pi/2, \\ -1, & \pi/2 < t < \pi. \end{cases}$$

(a) Interpret the boundary conditions physically.

(b) Find the solution $u(x,t) = X(x)T(t)$ of $u_{xx} = u_t$ in separated form Show all steps of the separation of variables process clearly.

(b) Find the solution $u(x,t)$ to the above heat problem. Write your answer in summation notation.

(c) Use the first three terms of your answer to (b) to find the temperature at the middle of the bar after 1 s.

4.5 The wave equation in one dimension

The theory of the vibrating string touches on musical theory and the theory of oscillating waves, so has likely been a concern of scholars since ancient times. Nevertheless, it wasn't until the late 1700s that mathematical progress was made. Although the problem of describing mathematically a vibrating string requires no calculus, the solution does. With the advent of calculus, Jean le Rond d'Alembert, Daniel Bernoulli, Leonard Euler, and Joseph-Louis Lagrange were able to arrive at solutions to the one-dimensional wave equation in the eighteenth century. Daniel Bernoulli's solution dealt with an infinite series of sines and cosines (derived from what we now call a "Fourier series," although it predates it); his contemporaries did not believe that he was correct. Bernoulli's technique would be later used by Joseph Fourier when he solved the thermodynamic heat equation in 1807. It is Bernoulli's idea that we discuss here as well. Euler was wrong: Bernoulli's method was basically correct after all.

Now, d'Alembert was mentioned in the section on the transport equation (§4.1.1) and it is worthwhile very briefly discussing what his basic idea was. The theorem of d'Alembert on the solution to the wave equation is stated roughly as follows: The partial differential equation

$$\frac{\partial^2 w}{\partial t^2} = c^2 \cdot \frac{\partial^2 w}{\partial x^2}$$

is satisfied by any function of the form $w = w(x,t) = g(x+ct)+h(x-ct)$, where g and h are "arbitrary" functions. (This is called "the d'Alembert solution.") Geometrically speaking, the idea of the proof is to observe that $\frac{\partial w}{\partial t} \pm c\frac{\partial w}{\partial x}$ is a constant times the directional derivative $D_{\vec{v_\pm}} w(x,t)$, where $\vec{v_\pm}$ is a unit vector in the direction $\langle \pm c, 1\rangle$. Therefore, you integrate

$$D_{\vec{v_-}} D_{\vec{v_+}} w(x,t) = (\text{const.})\frac{\partial^2 w}{\partial t^2} - c^2 \cdot \frac{\partial^2 w}{\partial x^2} = 0$$

twice, once in the $\vec{v_+}$ direction and once in the $\vec{v_-}$, to get the solution. Easier said than done, but still, that's the idea.

The wave equation with zero ends boundary conditions models the motion of a (perfectly elastic) guitar string of length L:

$$\begin{cases} c^2 \frac{\partial^2 w(x,t)}{\partial x^2} = \frac{\partial^2 w(x,t)}{\partial t^2} \\ w(0,t) = w(L,t) = 0. \end{cases} \tag{4.14}$$

Here $w(x,t)$ denotes the displacement from rest of a point x on the string at time t. The initial displacement $f(x)$ and initial velocity $g(x)$ at specified by the equations

$$w(x,0) = f(x), \qquad w_t(x,0) = g(x).$$

4.5.1 Method

The solution to the wave equation (4.14) can be derived using Fourier series, as with the heat equation.

- Find the sine series of $f(x)$ and $g(x)$:

$$f(x) \sim \sum_{n=1}^{\infty} b_n(f) \sin\left(\frac{n\pi x}{L}\right), \qquad g(x) \sim \sum_{n=1}^{\infty} b_n(g) \sin\left(\frac{n\pi x}{L}\right).$$

- The solution is

$$w(x,t) = \sum_{n=1}^{\infty} \left[b_n(f) \cos\left(c\frac{n\pi t}{L}\right) + \frac{L b_n(g)}{cn\pi} \sin\left(c\frac{n\pi t}{L}\right)\right] \sin\left(\frac{n\pi x}{L}\right). \tag{4.15}$$

Example 4.5.1. Let

$$f(x) = \begin{cases} -1, & 0 \le t \le \pi/2, \\ 2, & \pi/2 < t < \pi, \end{cases}$$

and let $g(x) = 0$. Then $L = \pi$, $b_n(g) = 0$, and

$$b_n(f) = \frac{2}{\pi} \int_0^{\pi} f(x) \sin(nx) dx = -2\,\frac{2\,\cos(n\pi) - 3\,\cos(\frac{1}{2}\,n\pi) + 1}{n}.$$

Thus

$$f(x) \sim b_1(f) \sin(x) + b_2(f) \sin(2x) + \cdots = \frac{2}{\pi} \sin(x) - \frac{6}{\pi} \sin(2x) + \frac{2}{3\pi} \sin(3x) + \ldots.$$

The function $f(x)$, and some of the partial sums of its sine series, looks like the plot in Figure 4.9.

This was computed using the following Sage commands.

Sage

```
sage: x = var("x")
sage: f1(x) = -1
sage: f2(x) = 2
sage: f = Piecewise([[(0,pi/2),f1],[(pi/2,pi),f2]])
sage: P1 = f.plot(rgbcolor=(1,0,0))
sage: b50 = [f.sine_series_coefficient(n,pi) for n in range(1,50)]
sage: ss50 = sum([b50[i-1]*sin(i*x) for i in range(1,50)])
sage: b50[0:5]
[2/pi, -6/pi, 2/3/pi, 0, 2/5/pi]
sage: P2 = ss50.plot(-5,5,linestyle="--")
sage: (P1+P2).show()
```

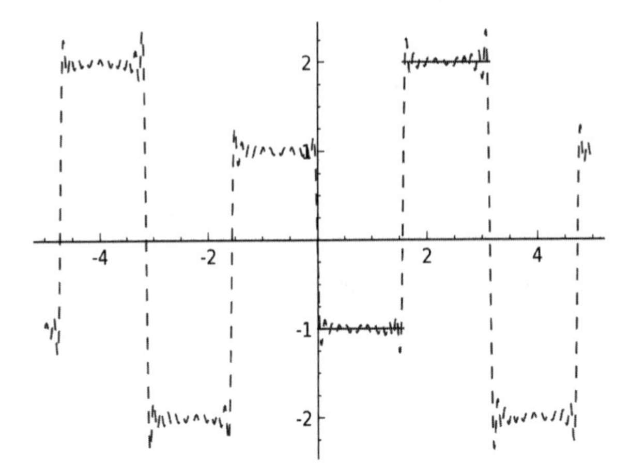

Figure 4.9: Using fifty terms of the sine series of $f(x)$.

As you can see, taking more and more terms gives functions that better and better approximate $f(x)$.

The solution to the wave equation, therefore, is (4.15). Taking only the first fifty terms of this series, the graph of the solution at $t = 0$, $t = 0.1$, $t = 1/5$, and $t = 1/4$, looks approximately like Figure 4.10.

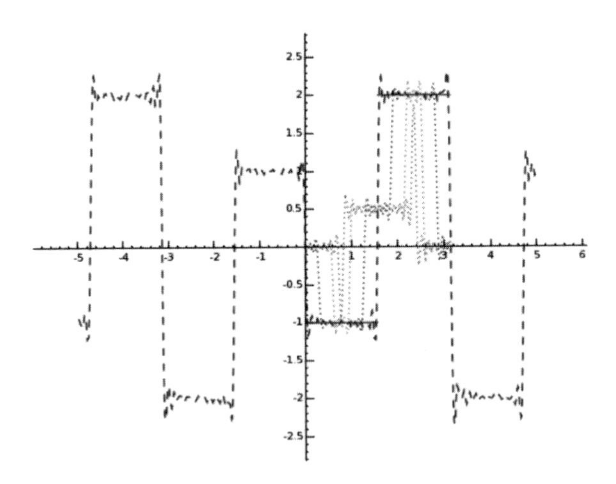

Figure 4.10: Wave equation with $c = 3$.

Figure 4.10 was produced using the following **Sage** commands.

```
Sage
sage: t = var("t")
sage: w50t1 = sum([b50[i-1]*sin(i*x)*cos(3*i*(1/10)) for i in range(1,50)])
sage: P3 = w50t1.plot(0,pi,linestyle=":")
```

```
sage: w50t2 = sum([b50[i-1]*sin(i*x)*cos(3*i*(1/5)) for i in range(1,50)])
sage: P4 = w50t2.plot(0,pi,linestyle=":",rgbcolor=(0,1,0))
sage: w50t3 = sum([b50[i-1]*sin(i*x)*cos(3*i*(1/4)) for i in range(1,50)])
sage: P5 = w50t3.plot(0,pi,linestyle=":",rgbcolor=(1/3,1/3,1/3))
sage: (P1+P2+P3+P4+P5).show()
```

Of course, taking terms would give a better approximation to $w(x,t)$. Taking the first one hundred terms of this series (but with different times) is given in Figure 4.11.

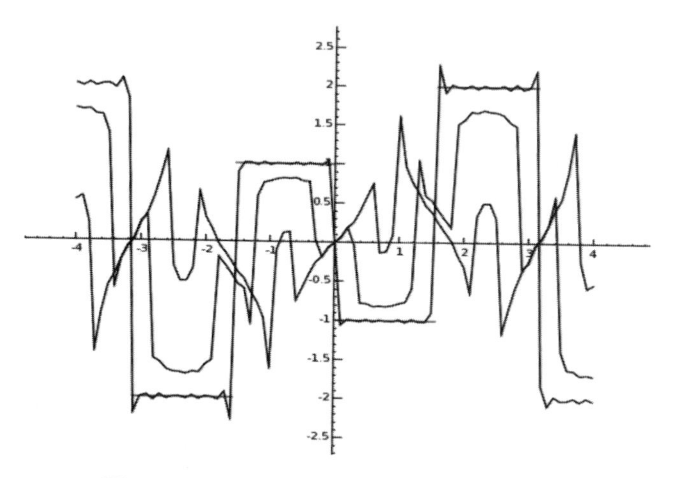

Figure 4.11: Wave equation with $c = 3$.

Exercises

1. Solve the wave equation

$$\begin{cases} 2\frac{\partial^2 w(x,t)}{\partial x^2} = \frac{\partial^2 w(x,t)}{\partial t^2} \\ w(0,t) = w(3,t) = 0 \\ w(x,0) = x \\ w_t(x,0) = 0, \end{cases}$$

using Sage to plot approximations as above.

2. Let

$$f(x) = \begin{cases} -1 & 0 < x < \pi/2 \\ 2 & \pi/2 < x < \pi. \end{cases}$$

   (a) Graph $f(x)$ over $0 < x < \pi$.

   (b) Compute the sine series of $f(x)$. Both

- write your answer in summation notation, and
- write down the sum of the first three nonzero terms.

(c) Find the value of the sine series of $f(x)$ at

- $x = 0$,
- $x = \pi/2$,
- $x = 2000\pi$,
- $x = 2001\pi$.

(d) Solve the wave equation with $\alpha = 3$, $u(x,0) = f(x)$, $u_t(x,0) = 0$, and zero ends.

4.6 The Schrödinger equation

The one-dimensional *Schrödinger equation* for a free particle is

$$ik\frac{\partial^2\psi(x,t)}{\partial x^2} = \frac{\partial\psi(x,t)}{\partial t},$$

where $k > 0$ is a constant (involving Planck's constant and the mass of the particle) and $i = \sqrt{-1}$ as usual. The solution ψ is called the *wave function* describing the instantaneous "state" of the particle. For the analog in three dimensions (which is the one actually used by physicists—the one-dimensional version we are dealing with is a simplified mathematical model), one can interpret the square of the absolute value of the wave function as the probability density function for the particle to be found at a point in space. In other words, $|\psi(x,t)|^2\,dx$ is the probability of finding the particle in the "volume dx" surrounding the position x, at time t.

If we restrict the particle to a "box" then (for our simplied one-dimensional quantum-mechanical model) we can impose a boundary condition of the form

$$\psi(0,t) = \psi(L,t) = 0,$$

and an initial condition of the form

$$\psi(x,0) = f(x), \qquad 0 < x < L.$$

Here f is a function (sometimes simply denoted $\psi(x)$) which is normalized so that

$$\int_0^L |f(x)|^2\,dx = 1.$$

If $|\psi(x,t)|^2$ represents a probability density function of finding a particle "at x" at time t then $\int_0^L |f(x)|^2\,dx$ represents the probability of finding the particle somewhere in the "box" initially, which is of course 1.

4.6.1 Method

- Find the sine series of $f(x)$:

$$f(x) \sim \sum_{n=1}^{\infty} b_n(f) \sin\left(\frac{n\pi x}{L}\right),$$

- The solution is

$$\psi(x,t) = \sum_{n=1}^{\infty} b_n(f) \sin\left(\frac{n\pi x}{L}\right) \exp\left(-ik\left(\frac{n\pi}{L}\right)^2 t\right).$$

Each of the terms

$$\psi_n(x,t) = b_n \sin\left(\frac{n\pi x}{L}\right) \exp\left(-ik\left(\frac{n\pi}{L}\right)^2 t\right).$$

is called a *standing wave* (although in this case sometimes b_n is chosen so that $\int_0^L |\psi_n(x,t)|^2\, dx = 1$).

Example 4.6.1. Let

$$f(x) = \begin{cases} -1, & 0 \le x \le 1/2, \\ 1, & 1/2 < x < 1. \end{cases}$$

Then $L = 1$ and

$$b_n(f) = \frac{2}{1}\int_0^1 f(x) \sin\left(\frac{n\pi x}{1}\right) dx = \frac{1}{n\pi}\left(-1 + 2\cos\left(\frac{n\pi}{2}\right) - \cos(n\pi)\right).$$

Thus

$$\begin{aligned} f(x) &\sim b_1(f)\sin(\pi x) + b_2(f)\sin(2\pi x) + \ldots \\ &= \textstyle\sum_n \frac{1}{n\pi}(-1 + 2\cos(\frac{n\pi}{2}) - \cos(n\pi)) \cdot \sin(n\pi x)\ . \end{aligned}$$

Taking more and more terms gives functions that better and better approximate $f(x)$. The solution to Schrödinger's equation, therefore, is

$$\psi(x,t) = \sum_{n=1}^{\infty} \frac{1}{n\pi}(-1 + 2\cos(n\pi/2) - \cos(n\pi)) \cdot \sin(n\pi x) \cdot \exp(-ik(n\pi)^2 t).$$

Explanation

Where does the solution in the above example come from? It comes from the method of separation of variables and the superposution principle. Here is a short explanation.

First, assume the solution to the partial differential equation $ik\frac{\partial^2\psi(x,t)}{\partial x^2} = \frac{\partial\psi(x,t)}{\partial t}$ has the "factored" form

$$\psi(x,t) = X(x)T(t)$$

for some (unknown) functions X, T. If this function solves the partial differential equation then it must satisfy $kX''(x)T(t) = X(x)T'(t)$, or

$$\frac{X''(x)}{X(x)} = \frac{1}{ik}\frac{T'(t)}{T(t)}.$$

Since x, t are independent variables, these quotients must be constant. In other words, there must be a constant C such that

$$\frac{T'(t)}{T(t)} = ikC, \quad X''(x) - CX(x) = 0.$$

Now we have reduced the problem of solving the one partial differential equation to two ODEs (which is good), but with the price that we have introduced a constant that we don't know, namely C (which maybe isn't so good). The first ODE is easy to solve:

$$T(t) = A_1 e^{ikCt}$$

for some constant A_1. It remains to "determine" C.

Case $C > 0$. Write (for convenience) $C = r^2$, for some $r > 0$. The ODE for X implies $X(x) = A_2\exp(rx) + A_3\exp(-rx)$, for some constants A_2, A_3. Therefore

$$\psi(x,t) = A_1 e^{-ikr^2t}(A_2\exp(rx) + A_3\exp(-rx)) = (a\exp(rx) + b\exp(-rx))e^{-ikr^2t},$$

where A_1A_2 has been renamed a and A_1A_3 has been renamed b. This will not match the boundary conditions unless a and b are both 0.

Case $C = 0$. This implies $X(x) = A_2 + A_3x$, for some constants A_2, A_3. Therefore

$$\psi(x,t) = A_1(A_2 + A_3x) = a + bx,$$

where A_1A_2 has been renamed a and A_1A_3 has been renamed b. This will not match the boundary conditions unless a and b are both 0.

Case $C < 0$. Write (for convenience) $C = -r^2$ for some $r > 0$. The ODE for X implies $X(x) = A_2\cos(rx) + A_3\sin(rx)$ for some constants A_2, A_3. Therefore

$$\psi(x,t) = A_1 e^{-ikr^2t}(A_2\cos(rx) + A_3\sin(rx)) = (a\cos(rx) + b\sin(rx))e^{-ikr^2t},$$

where A_1A_2 has been renamed a and A_1A_3 has been renamed b. This will not match the boundary conditions unless $a = 0$ and $r = \frac{n\pi}{L}$

These are the solutions of the heat equation which can be written in factored form. By superposition, the general solution is a sum of these:

$$\psi(x,t) = \sum_{n=1}^{\infty}(a_n\cos(r_nx) + b_n\sin(r_nx))e^{-ikr_n^2t} = b_1\sin(r_1x)e^{-ikr_1^2t} + b_2\sin(r_2x)e^{-ikr_2^2t} + \ldots, \tag{4.16}$$

for some b_n, where $r_n = \frac{n\pi}{L}$. Note the similarity with Fourier's solution to the heat equation.

There is one remaining condition which our solution $\psi(x,t)$ must satisfy. We have not yet used the initial condition $\psi(x,0) = f(x)$. We do that next.

Plugging $t = 0$ into (4.16) gives

$$f(x) = \psi(x,0) = \sum_{n=1}^{\infty} b_n \sin\left(\frac{n\pi}{L}x\right) = b_1 \sin\left(\frac{\pi}{L}x\right) + b_2 \sin\left(\frac{2\pi}{L}x\right) + \dots .$$

In other words, if $f(x)$ is given as a sum of these sine functions, or if we can somehow express $f(x)$ as a sum of sine functions, then we can solve Schrödinger's equation. In fact there is a formula for these coefficients b_n:

$$b_n = \frac{2}{L}\int_0^L f(x)\cos\left(\frac{n\pi}{L}x\right)dx.$$

It is this formula that is used in the solutions above.

Bibliography

[A-ns] David Acheson, *Elementary Fluid Dynamics* (Oxford: Oxford University Press, 1990).

[A-ode] Kendall Atkinson, Weimin Han, Laurent Jay, and David W. Stewart, *Numerical Solution of Ordinary Differential Equations* (New York: John Wiley and Sons, 2009).

[A-pde] Wikipedia articles on the transport equation:
`http://en.wikipedia.org/wiki/Advection`
`http://en.wikipedia.org/wiki/Advection_equation`

[A-uc] Wikipedia entry for the annihilator method:
`http://en.wikipedia.org/wiki/Annihilator_method`

[BD-intro] W. Boyce and R. DiPrima, *Elementary Differential Equations and Boundary Value Problems*, 8th ed. (New York: John Wiley and Sons, 2005).

[B-fs] Wikipedia entry for Daniel Bernoulli:
`http://en.wikipedia.org/wiki/Daniel_Bernoulli`

[B-ps] Wikipedia entry for the Bessel functions:
`http://en.wikipedia.org/wiki/Bessel_function`

[B-rref] Robert A. Beezer, *A First Course in Linear Algebra*, released under the GNU Free Documentation License, available at `http://linear.ups.edu/` and `http://www.lulu.com/`.

[BS-intro] General wikipedia introduction to the Black-Scholes model:
`http://en.wikipedia.org/wiki/Black-Scholes`

[C-ivp] General wikipedia introduction to the catenary:
`http://en.wikipedia.org/wiki/Catenary`

[C-linear] General wikipedia introduction to RLC circuits:
`http://en.wikipedia.org/wiki/RLC_circuit`

[CS-rref] Wikipedia article on normal modes of coupled springs: `http://en.wikipedia.org/wiki/Normal_mode`

[D-df] Wikipedia introduction to direction fields:
http://en.wikipedia.org/wiki/Slope_field

[D-spr] Wikipedia entry for damped motion: http://en.wikipedia.org/wiki/Damping

[E-num] General wikipedia introduction to Euler's method: http://en.wikipedia.org/wiki/Euler_integration

[Eu1-num] Wikipedia entry for Euler: http://en.wikipedia.org/wiki/Euler

[Eu2-num] MacTutor entry for Euler:
http://www-groups.dcs.st-and.ac.uk/~history/Biographies/Euler.html

[F1-fs] Wikipedia Fourier series article
http://en.wikipedia.org/wiki/Fourier_series

[F2-fs] MacTutor Fourier biography:
http://www-groups.dcs.st-and.ac.uk/%7Ehistory/Biographies/Fourier.html

[F-1st] General wikipedia introduction to first-order linear differential equations:
http://en.wikipedia.org/wiki/Linear_differential_equation#First_order_equation

[Gl] Andrew Gleason, *Fundamentals of Abstract Analysis*, Reading, MA: Addison-Wesley, 1966; corrected reprint, Boston: Jones and Bartlett, 1991.

[H-fs] General wikipedia introduction to the heat equation:
http://en.wikipedia.org/wiki/Heat_equation

[H-intro] General wikipedia introduction to Hooke's law: http://en.wikipedia.org/wiki/Hookes_law

[H-ivp] General wikipedia introduction to the hyperbolic trigonometric function:
http://en.wikipedia.org/wiki/Hyperbolic_function

[H-rref] Jim Hefferon, *Linear Algebra*, released under the GNU Free Documentation License, available at http://joshua.smcvt.edu/linearalgebra/ and http://www.createspace.com.

[HSD] Morris W. Hirsch, Stephen Smale, and Robert L. Devaney, *Differential Equations, Dynamical Systems and An Introduction to Chaos*, 2nd ed. (Amsterdam: Elsevier Academic Press, 2004).

[H-sde] Desmond J. Higham, "An Algorithmic Introduction to Numerical Simulation of Stochastic Differential Equations," *SIAM Review*, Vol. 43, No. 3 (2001), pp. 525-546.

[H1-spr] Wikipedia entry for Robert Hooke: http://en.wikipedia.org/wiki/Robert_Hooke

[H2-spr] MacTutor entry for Hooke:
http://www-groups.dcs.st-and.ac.uk/%7Ehistory/Biographies/Hooke.html

[KL-cir] Wikipedia entry for Kirchoff's laws:
http://en.wikipedia.org/wiki/Kirchhoffs_circuit_laws

[K-cir] Wikipedia entry for Kirchoff: http://en.wikipedia.org/wiki/Gustav_Kirchhoff

[LA-sys] Lanchester automobile information:
http://www.amwmag.com/L/Lanchester_World/lanchester_world.html

[La-sys] Frederick William Lanchester, *Aviation in Warfare: The Dawn of the Fourth Arm*, Constable and Co., London, 1916.

[L-intro] F. W. Lanchester, *Mathematics in Warfare*, in *The World of Mathematics*, J. Newman ed., vol. 4, pp.138-157. (New York: Simon and Schuster 1956, reprinted New York: Dover 2000).

[L-linear] General wikipedia introduction to linear independence:
http://en.wikipedia.org/wiki/Linearly_independent

[L-lt] Wikipedia entry for Laplace: http://en.wikipedia.org/wiki/Pierre-Simon_Laplace

[Lo-intro] General wikipedia introduction to the logistic function model of population growth:
http://en.wikipedia.org/wiki/Logistic_function

[L-sys] Wikipedia entry for Lanchester:
http://en.wikipedia.org/wiki/Frederick_William_Lanchester

[LT-lt] Wikipedia entry for Laplace transform:
http://en.wikipedia.org/wiki/Laplace_transform

[L-var] Wikipedia article on Joseph Louis Lagrange:
http://en.wikipedia.org/wiki/Joseph_Louis_Lagrange

[M] Maxima, a general purpose computer algebra system.
http://maxima.sourceforge.net/

[M-fs] Wikipedia entry for the physics of music:
http://en.wikipedia.org/wiki/Physics_of_music

[M-intro] Niall J. MacKay, "Lanchester Combat Models," May 2005. http://arxiv.org/abs/math.HO/0606300

[M-mech] General wikipedia introduction to Newtonian mechanics:
http://en.wikipedia.org/wiki/Classical_mechanics

[M-ps] Sean Mauch, "Introduction to Methods of Applied Mathematics," http://www.its.caltech.edu/~sean/book/unabridged.html

[M-zom] P. Munz, I. Hudea, J. Imad, R. Smith?, "When Zombies Attack! Mathematical Modelling of an Outbreak of Zombie Infection," In: *Infectious Disease Modelling Research Progress, 2009*, pp. 133-150. (New York: Nova Science, 2009).

[N-cir] Wikipedia entry for electrical networks: http://en.wikipedia.org/wiki/Electrical_network

[N-intro] David H. Nash, "Differential equations and the Battle of Trafalgar," *The College Mathematics Journal*, Vol. 16, No. 2 (Mar., 1985), pp. 98-102.

[N-mech] General wikipedia introduction to Newton's three laws of motion: http://en.wikipedia.org/wiki/Newtons_Laws_of_Motion

[NS-intro] General wikipedia introduction to Navier-Stokes equations: http://en.wikipedia.org/wiki/Navier-Stokes_equations
Clay Math Institute prize page:
http://www.claymath.org/millennium/Navier-Stokes_Equations/

[O-ivp] General wikipedia introduction to the harmonic oscillator: http://en.wikipedia.org/wiki/Harmonic_oscillator

[P-fs] Howard L. Penn, "Computer Graphics for the Vibrating String," *The College Mathematics Journal*, Vol. 17, No. 1 (Jan., 1986), pp. 79-89.

[P-intro] General wikipedia introduction to the Peano existence theorem: http://en.wikipedia.org/wiki/Peano_existence_theorem

[PL-intro] General wikipedia introduction to the Picard existence theorem: http://en.wikipedia.org/wiki/Picard-Lindelof_theorem

[Post] Emil L. Post, "Generalized Differentiation," Transactions of the American Mathematical Society, Vol. 32 (1930), pp. 723-781.

[P1-ps] Wikipedia entry for power series: http://en.wikipedia.org/wiki/Power_series

[P2-ps] Wikipedia entry for the power series method: http://en.wikipedia.org/wiki/Power_series_method

[R-cir] General wikipedia introduction to LRC circuits: http://en.wikipedia.org/wiki/RLC_circuit

[Rikitake] T. Rikitake, "Oscillations of a System of Disk Dynamos," Proceedings of the Royal Cambridge Philosophical Society, 54 (1958), pp. 89-105.

[R-ps] Wikipedia entry for the recurrence relations: http://en.wikipedia.org/wiki/Recurrence_relations

[SH-spr] Wikipedia entry for simple harmonic motion: http://en.wikipedia.org/wiki/Simple_harmonic_motion

[S-intro] William Stein and the Sage Developers Group, Sage: *Mathematical software*, version 4.8:
http://www.sagemath.org/

[S-pde] W. Strauss, *Partial Differential Equations, an Introduction* (New York: John Wiley, 1992).

[U-uc] General wikipedia introduction to undetermined coefficients: http://en.wikipedia.org/wiki/Method_of_undetermined_coefficients

[V-var] Wikipedia introduction to variation of parameters:
http://en.wikipedia.org/wiki/Method_of_variation_of_parameters

[W-intro] General wikipedia introduction to the wave equation: http://en.wikipedia.org/wiki/Wave_equation

[W-linear] General wikipedia introduction to the Wronskian
http://en.wikipedia.org/wiki/Wronskian

[W-mech] General wikipedia introduction to Wile E. Coyote and the RoadRunner:
http://en.wikipedia.org/wiki/Wile_E._Coyote_and_Road_Runner

[Wr-linear] St. Andrews MacTutor entry for Wronski
http://www-groups.dcs.st-and.ac.uk/history/Biographies/Wronski.html

Index